S0-AIZ-907

THE ART AND
SCIENCE OF RADIO

Linda Busby

University of South Alabama

Donald Parker

Vice President of Research and Development

Quality Micro Systems (QMS)

ALLYN AND BACON, INC.

Boston   London   Sydney   Toronto

Consulting Editor: Robert Avery
Series Editor: Bill Barke

Copyright © 1984 by Allyn and Bacon, Inc., 7 Wells Avenue, Newton, Massachusetts 02159. All rights reserved. No part of the material protected by this copyright notice may be reproduced or utilized in any form or by any means, electronic or mechanical, including photocopying, recording, or by any information storage or retrieval system, without written permission from the copyright owner.

Library of Congress Cataloging in Publication Data

Busby, Linda J., 1947–
    The art and science of radio.

    Includes bibliographies and index.
    1. Radio broadcasting—United States.   2. Radio
industry and trade—United States.   3. Radio—Apparatus and
supplies.   4. Radio plays—Production and direction.
5. Radio stations—Administration.   I. Parker, Donald L.
II. Title.
PN1991.3.U6B76   1984   384.54′0973   83-25742
ISBN 0-205-08049-9

Printed in the United States of America

10   9   8   7   6   5   4   3   2   1   89   88   87   86   85   84

384.54
B97

86-2183

To
Betty Jean Busby
Louie Busby
Lois Parker
and Orin Harvey Parker

CONTENTS

Chapter Four

BROADCAST COPYWRITING   83

Chapter Five

THE RADIO NEWS PROCESS   99

Chapter Six

RADIO PROGRAMMING   116

Appendix A

THE METRIC SYSTEM

Appendix B

ALL ABOUT DECIBELS

PREFACE

As the twentieth century is characterized by the development of technology, so the development and progression of radio are on track with the twentieth century.

Marconi sent a wireless signal across his father's estate in Italy in 1895. The first commercially licensed station went on the air in 1920. National networks made debuts in 1926 and 1927. Radio nurtured Americans during the depression and informed Americans during World War II. Radio nearly died as a national medium during the late 1950s and early 1960s, but soon was reborn with the DJ and audience-participation promotions. In the 1980s radio is stronger than ever. The audience has increased, the number of stations has increased, the number of networks has increased. Satellites, digital audio, and AM stereo will provide a higher quality signal for the listener.

The National Association of Broadcasters reports that radio provides most of the entry-level professional positions in all broadcasting and accounts for more than 116,000 full-time jobs in the broadcasting industry. In recent years, students have developed again an interest in radio. Some students are interested in every aspect of the industry, while others want to learn audio production techniques, which can be applied to a variety of career fields.

The Art and Science of Radio will serve both the students who want to explore every aspect of broadcasting and the students who want to learn audio production. The Art and Science of Radio is comprehensive. The textbook provides information on the history of radio, the science of radio, audio production, broadcast copywriting, radio journalism, radio station ownership and operations, sales and promotion, broadcast regulations, and the future of radio in the United States.

Several aspects of The Art and Science of Radio should make it particularly appealing to students. The "Questions for Review and Discussion" at the close of each chapter will help focus students' attention on the key information in each chapter. Another important feature of this book is the large number of drawings, illustrations, and photographs. The scientific aspects of radio and the technology of production are amply illustrated.

Several aspects of *The Art and Science of Radio* should make it particularly appealing to instructors. The "Questions for Review and Discussion" will help teachers focus classroom discussion and also provide excellent examination material. The *Instructor's Manual for the Art and Science of Radio* provides course outlines, chapter summaries, lecture topics, and activities for student participation. The comprehensive nature of the textbook lends itself to a variety of course structures.

With broadcast regulations and technologies changing rapidly, many sections of the text were difficult to write. In those sections of the text dealing with regulations and technologies that are likely to change in the near future, the reader is advised to check professional journals for updates.

Compiling the materials for this text has taken several years. The task would not have been possible without help. We are especially grateful to our editor, Bill Barke at Allyn and Bacon, to Judy Fiske at Allyn and Bacon, and to Sarah Doyle at Bywater Production Services. We are also grateful for the assistance that we received from Bob Avery at the University of Utah.

In addition to individuals, a number of institutions provided photographs and illustrations. We appreciate the radio stations and equipment manufacturers that responded positively to our requests.

We hope *The Art and Science of Radio* is a contribution to the pedagogic literature in the broadcasting curriculum.

L. J. B.

D. L. P.

Chapter One

RADIO: YESTERDAY
AND TODAY

*Radio is the miracle of the ages. Aladdin's Lamp, the Magic
Carpet, and the Seven League Boots of fable and every vision
that mankind has ever entertained . . . of laying hold upon the
Almighty pale into insignificance beside the accomplished
fact of radio.*

> Gleason L. Archer,
> *History of Radio,* 1938

Radio, which started in America as a recreational activity for amateurs eager to
have their signals heard as far as next door, became a multimillion dollar industry in a very short space of time. When Guglielmo Marconi experimented
with wireless telegraphy in his native Italy in 1895, he was primarily concerned
with a very limited form of communication that would result in a more
economical and faster method of sending Morse code, but the concept of
wireless developed so rapidly that by 1925, Secretary of Commerce Herbert
Hoover called radio "a vital force in American life." To understand fully the impact of radio on Americans, and to understand radio's future, requires a look at
radio's past. This chapter explores radio's beginnings and its development into
the vast communication system that we know today.

THE DEVELOPMENT OF RADIO

The "Father of Radio"

No invention, discovery, or new body of knowledge pops into existence overnight. Every step forward in human understanding follows a series of investiga-
tions that spark additional research and add to existing knowledge. The
development of radio is no exception.

1

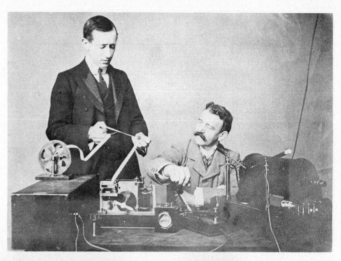

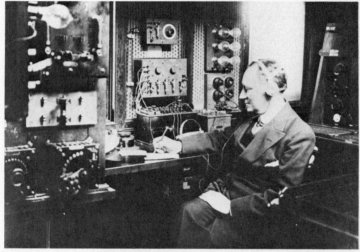

Exhibit 1.1. Marconi and His Wireless Devices
(*opposite, top*) Marconi with his associate George Kemp. The two experimenters
posed for the Biograph Company which was making a film about wireless (1901).
(Photo courtesy of the Marconi Company Limited, Marconi House, Chelmsford,
Essex.) (*middle*) Marconi aboard naval laboratory, *Elettra*. Marconi was a world
leader in naval and ship-to-shore communication experiments. (Photo courtesy
of the Marconi Company Limited, Marconi House, Chelmsford, Essex.) (*bottom*)
Marconi, twenty-two years old, displayed the mysterious black box containing a
coherer, tapper, and a relay (1896). (Photo courtesy of the Marconi Company
Limited, Marconi House, Chelmsford, Essex.)

Marconi has been called the father of radio. As a teenager he experimented
with radio waves, and by 1895 he was sending radio signals across short
distances at the family estate in Italy. The Italian government told the young in-
ventor it was not interested in his experimentation. So the Marconi family
shipped the young man and his wireless invention off to England where he was
placed in contact with government officials and patent attorneys. By 1897,
Marconi had received a patent on his wireless and, shortly thereafter, he
formed the Marconi Telegraph Company Ltd. Marconi was only twenty-three
years old at the time but was well on his way to international acclaim and personal wealth. His patent and his company inaugurated the world into the age of
radio communication. Exhibit 1.1 shows Marconi with some of his wireless
devices.

Marconi envisaged radio, not as a mass medium of communication, but as a
medium for sending Morse code. Since the British had the largest navy in the
world, and the navy needed a communication system to unite the ships at sea,
Marconi began experiments designed to attract the British Navy. Newspapers
around the world carried accounts of his experiments, and military observers
came from around the globe to watch Marconi at work. While Marconi had
begun to use the wireless for telegraphy, wireless had a long way to go to
become the communication system we know today.

Radio's Development in America

Some Americans had experimented with wireless as early as 1866. These early
American radio enthusiasts were much more localized in their renown and
limited in their experimentations than was Marconi.

Nathan B. Stubblefield of Murray, Kentucky, is rumored to have transmitted voice via radio as early as 1892.[1] Stubblefield may have been a genius of the
calibre of Thomas Edison or Marconi, but he lacked the sense of public relations that kept the names of Edison and Marconi in the limelight. When members of the press or visitors from the business world journeyed to Murray,

---

1. Erik Barnouw, *A History of Broadcasting in the United States,* vol. 1, *A Tower in Babel* (New York:
Oxford University Press, 1966), pp. 18–19.

Exhibit 1.2. Nathan Stubblefield and His Inventions
The Stubblefield family gathered at the home of Nathan B. Stubblefield, Murray,
Kentucky. The inventions and the newspaper articles about the work of Nathan
Stubblefield are proudly displayed. Nathan Stubblefield and his wife are at the far
right of the picture (numbers 13 and 14). The Stubblefield children are on the porch.
(Photo courtesy of Murray State University Special Collections Library, Murray,
Kentucky.)

Kentucky, to inspect Stubblefield's inventions, they were frequently sent packing by threats from a shotgun. Stubblefield had offers from capitalists and
financiers, but he refused them all, wanting to perfect his inventions before
marketing them. Also, he feared that someone would steal his ideas. He died of
starvation in an isolated hut in the Kentucky woods.[2] (See Exhibit 1.2.)

Two Americans who became internationally renowned for their work with
radio are Lee de Forest and Reginald Fessenden. De Forest was granted over
two hundred radio patents, and Fessenden was responsible for the first voice
transmission via radio. (See Exhibit 1.3.)

Wireless had been used for sending Morse code by interrupting the radio
wave signal and transmitting a series of short electrical bursts. In an historic
broadcast on Christmas Eve of 1906, Fessenden accomplished a broadcasting
first by sending an uninterrupted radio wave on which he superimposed a
human voice by using a telephone microphone. The first real radio broadcast is

2.   For more information on Stubblefield see Harvey Geller, "Nathan Stubblefield: The Radio
     Prophet of the Kentucky Fields," *High Fidelity,* November 1977, pp. 79–83.

Exhibit 1.3. Reginald Fessenden at His Laboratory
(*top*) The Brant Rock operators. Reginald Fessenden is seated in the middle. (*bottom*) Reginald Fessenden, 1905. (Photos from *Radio's First Voice,* by Ormond Raby. Reprinted by permission of the author.)

attributed to this Canadian-American, who prepared a miniprogram with a Christmas theme to be transmitted from his laboratory at Brant Rock, Massachusetts. Accounts of the broadcast indicate its significance in the preradio world of 1906.

> Early that evening wireless operators on ships within a radius of several hundred miles sprang to attention as they caught the call, "CQ, CQ," in Morse code. Was it a ship in distress? They listened eagerly and, to their amazement, heard a human voice coming from their instruments—someone speaking! Then a woman's voice rose in song. It was uncanny! Many of them called to officers to come and listen; soon wireless rooms were crowded. Next someone was heard reading a poem. Then there was a violin solo; then a man made a speech and they could catch most of the words. Finally, everyone who had heard the program was asked to write to R. A. Fessenden at Brant Rock, Massachusetts—and many operators did.[3]

In 1906, de Forest introduced another radio development—an improved vacuum tube that he called an audion tube. (See Exhibit 1.4.) The audion tube could now control powerful currents and could amplify voice transmissions while also improving the clarity of reception. The audion addition to radio technology greatly improved voice transmission and was the essential development for modulating, amplifying, and decoding radio waves. By 1906, de Forest and Fessenden had earned their landmarks in radio history. These Americans and others were contributing to the technological progression in the development of radio.

Radio Becomes a Scientific Toy

Wireless devices were demonstrated at fairgrounds and department stores and attracted the attention of radio enthusiasts called "amateurs." These new enthusiasts would become the target audience for the sale of radio parts and equipment. By 1912, the radio amateurs operated home stations in communities across the country; the amateurs were self-regulated, operating when and on whatever frequencies they chose. Many times it was impossible to receive a clear signal because another amateur was operating on the same frequency. Radio was, for the amateur, a toy, an amusement.

In 1912, an event changed the status of radio from a toy to a serious means of life-saving communication. The event was the sinking of the great ship *Titanic*. Telegraphic word that the *Titanic* was sinking was received by twenty-one-year-old David Sarnoff, who manned the wireless station in the Wanamaker store in New York.

Sarnoff heard a signal: "S.S. *Titanic* ran into iceberg. Sinking fast." These

---

3.   Alvin F. Harlow, *Old Wires and New Waves* (New York: Appleton-Century, 1936), pp. 455–456. (Reprinted by Arno Press, 1971.)

Exhibit 1.4. Lee de Forest with His Audion Tube
Lee de Forest with a replica of the audion tube that he patented in 1906. This photograph was taken on his eighty-fourth birthday, August 26, 1957. He died in August 1961. In addition to the audion patent, he was granted over 200 radio patents during the years of his experimentation. (Photo from the Broadcast Pioneers Library Collection, Washington, D.C.)

coded letters started a chain of events that kept young Sarnoff at his post for a seventy-two-hour shift.

Seven hundred passengers were rescued, while over fifteen hundred were lost at sea. The *Titanic* disaster was not only responsible for changing the status of radio from a toy to a life-saving means of communication, but it was also responsible for radio legislation passed by Congress requiring the presence of ship radios and active ship operators on passenger vessels. This new legislation meant an even greater market for the radio companies that were developing into communication giants.

Exhibit 1.5. David Sarnoff, American Broadcast Visionary
(*above*) David Sarnoff in 1907 when he was an office boy at Marconi Wireless
Telegraph Company of America. (*opposite, top*) Sarnoff, who taught himself Morse
Code, was given the job as wireless operator for the station on Nantucket Island.
(*opposite, bottom*) Sarnoff began his career in the communication industry as an
office boy and went forward to become president of RCA and one of the major con-
tributors to modern mass communication. (Photos courtesy of RCA.)

Radio Becomes a Corporate Institution

During World War I, radio became the official property of the United States
Navy. From 1914 to 1918, amateurs were forbidden to use their radio equipment for fear that enemy forces would intercept crucial information and that
amateur transmissions would interfere with naval communications. Radio experimentation, however, did not cease during this period; it actually increased
since the navy gave experimenters a chance to demonstrate their radio inventions.

When the Navy finally relinquished control of the U.S. wireless stations in
1920, American Marconi was the only company with the capital and organizational expertise necessary to capture the U.S. radio market. The United States,
with the active participation of the Navy, began a series of negotiations to accomplish two important feats: 1) to put together an American radio enterprise
controlled primarily by United States interests; 2) to settle some of the patent
disputes so that the development of radio in America could proceed.

On November 20, 1919, a new enterprise emerged — the Radio Corporation
of America (RCA), the new company taking over the operations and assets of
American Marconi. David Sarnoff, the young wireless operator who had connected the United States with ships rescuing passengers from the sinking
Titanic, became the commercial manager of RCA. Ultimately, David Sarnoff
would become a major name in American broadcasting. (See Exhibit 1.5.)

THE DAWN OF COMMERCIAL BROADCASTING IN AMERICA

The First Licensed Station

At the close of World War I, Dr. Frank Conrad, a radio engineer at Westinghouse, continued his research and improved his station 8XK, located in Pittsburgh. Conrad broadcast regularly at 7:30 on Wednesday and Saturday evenings for a period of two hours. Assisted by his two sons, Conrad played phonograph records he obtained from local music stores.

Since Conrad worked for Westinghouse, the company took an interest and decided to finance the station, which went on the air November 2, 1920. KDKA's first broadcast was the Harding-Cox presidential election returns. The station's format included political news, music, and sports reports. Westinghouse expected no monetary return from the station but did expect KDKA to bring its parent company favorable publicity. Indeed, KDKA was placing Westinghouse in the national limelight. (See Exhibit 1.6.)

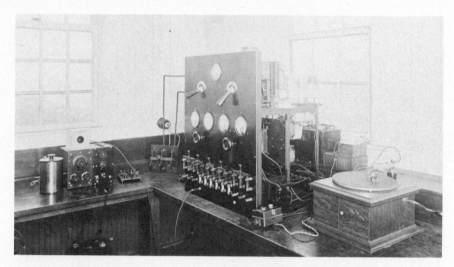

Exhibit 1.6. KDKA, America's First Licensed Broadcast Station
(*above*) Original broadcast transmitter of KDKA, Pittsburgh. This transmitter, using 50-watt tubes, was used November 2, 1920, when KDKA broadcast the Harding-Cox election returns. (*opposite, top*) Aeriola Jr. was the first popular-priced home radio receiver. This tiny crystal set, designed by KDKA engineers in 1921, was 6" × 6" × 7" in size, employed earphones, had a range of 12–15 miles, and sold for $25. At last, there was a set simple enough to be operated by nontechnical fans and inexpensive enough to be afforded by every household. (*opposite, bottom*) Carried aloft by a big balloon, an experimental antenna is tested in the Pittsburgh area by KDKA radio personnel. (Photos courtesy of Group W, Westinghouse Broadcasting Company.)

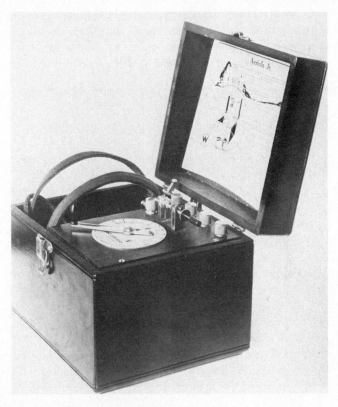

Other stations vie with KDKA for the honor of being the first regular broadcasting station in the United States. But KDKA has one piece of documentation with which the others cannot quarrel — it is the first commercially licensed standard broadcast station listed in the United States Department of Commerce records.

As stations around the country took to the air, the American public became even more enthusiastic. The first issue of *Radio Broadcast,* May 1922, provides a sense of the excitement that radio was generating:

> The rate of increase in the number of people who spend at least part of their evening listening in is almost incomprehensible. To those who have recently tried to purchase receiving equipment, some idea of the increase has undoubtedly occurred as they stood perhaps in the fourth or fifth row at the radio counter that they might place an order and it would be filled when possible. . . . The movement is probably not yet even at its height. It is still growing in some kind of geometric progression. . . . It seems quite likely that before the movement has reached its height, before the market for receiving apparatus becomes approximately saturated, there will be at least five million sets in this country.[4]

As more and more stations took to the air, more radio advertising was heard. Companies sponsored programs that bore the company name and sponsored musical groups who were living advertisements — A&P Gypsies; Ipana Troubadours; Lucky Strike Hour. Business began to realize the benefits of radio advertising. The early advertisements were called "indirect"; they did not plug a particular product but, rather, attempted to establish goodwill for the company in the public mind. In the "indirect" ads, the company name was mentioned and the company reminded the audience of its quality product and of the company history of quality service. Radio was considered to be a guest in the parlor; direct advertising would not have been polite.

Even in radio's early days there were protests against advertising. The sentiment of those individuals annoyed with advertising clutter on radio was expressed as early as November 1922 in a feature story carried by *Radio Broadcast.*

> Driblets of advertising . . . are floating through the ether every day. Concerts are seasoned here and there with a dash of advertising paprika. You can't miss it; every little classic number has a slogan all its own. . . . More of this sort of thing may be expected. And once the avalanche gets a good start, nothing short of an act of Congress or a repetition of Noah's excitement will suffice to stop it.[5]

4. *Radio Broadcast,* "Radio Currents," May 1922, pp. 1–2. It is interesting to note how far this author underestimated the saturation level for radio. As of 1984, there were over 413 million radio receivers in use in the United States.
5. Joseph H. Jackson, "Should Radio Be Used for Advertising?" *Radio Broadcast,* November 1922, p. 76.

Indeed, commercial broadcasting had come to America to stay. Amid some loud outcries, advertising would continue and would be the driving force in developing our vast system of commercial broadcasting.

The Development of Broadcasting Networks

Chain broadcasting—the process of connecting two or more stations for the same program—had been discussed since the earliest days of broadcasting. On a number of occasions, two or more stations were linked for simultaneous broadcast. But chain broadcasting really linked the nation when RCA formed a broadcast network—the National Broadcasting Company (NBC). The new network broadcast its first program on November 15, 1926, with a four-and-a-half-hour coast-to-coast hookup of an estimated twenty-five stations. The broadcast featured Walter Damrosch conducting the New York Symphony Orchestra.

Actually, RCA had two networks—NBC Blue, originating from WJZ (Newark, New Jersey), and NBC Red, originating from WEAF (New York). NBC had little difficulty programming for the two networks since its prestige and power attracted artists of the highest calibre.

The situation was not nearly so easy for America's second national network. Arthur Judson, a violinist, had changed his trade to artist management, and he wanted to provide artists for NBC. When NBC decided not to use Judson's management company, Judson joined with George A. Coats to form a new network. The difficulty experienced by the two men cannot be underestimated. They had difficulty signing stations and great difficulty paying the numerous bills associated with forming a new network. In January 1927, the network was finally formed and was named United Independent Broadcasters.

Because the financial burden of developing a network was so great, Judson was eager to get new investors and to sell part of his interests. In this effort, he joined with Columbia Phonograph Company to form the Columbia Phonograph Broadcasting System, and the new network tried desperately to find sponsors, but few appeared. The situation seemed almost hopeless, when a buyer for the network appeared.

The Paley family owned the Congress Cigar Company. When cigar sales were very low, young William Paley, twenty-six years old and a company executive, purchased some radio time to advertise La Palina cigars. The radio advertising brought immediate revenues into the cigar business, and the family was impressed with the power of radio and with the marketing skills of Bill Paley. When Judson and Coats sought a major shareholder for the new network, the Paley family stepped forward; the network gained the much-needed financial support and young Bill Paley became president of the Columbia Broadcasting System (CBS). Paley, who became a leader in the new broadcasting industry, was to establish policy at CBS for over half a century.

In this two-year period (1926–1927), two national networks had been formed—NBC and CBS. These networks molded the shape of broadcasting in the United States. The mainstay of the networks' programming was musical performances by talent who performed in posh studios. But, by 1929, drama too was entrenched in the program selections; NBC aired "Great Moments in History," "Biblical Dramas," and the "Eveready Hour," while CBS broadcast "Main Street" and "True Story." Other programs included some news and sports coverage plus coverage of special events.

Other networks offered programs to compete with NBC and CBS. The most significant of these was the Mutual Network, which started when station WGN (Chicago), WOR (New York), and WLW (Cincinnati) agreed to exchange programs and jointly sell time. The Mutual system was a group of independent stations that shared programs with each other but did not assume the restrictions of network affiliation that went with an NBC or a CBS contract. The most popular program produced by the Mutual system was "The Lone Ranger," produced at station WXYZ in Detroit. The Mutual Broadcasting System, organized in 1934, boasted over five hundred stations by 1948. Most of these were in the smaller markets, since NBC and CBS had signed up the major market stations. Mutual Broadcasting System survived the traumas of networking and is still an important force in the industry today.

The American Broadcasting Company (ABC) was formed in 1943 when the FCC adopted its "Chain Broadcasting Regulations," stating that a station could not be owned by a parent company having more than one network. This ruling meant that NBC, with its Red and Blue networks, was forced to divest itself of one of them. NBC sold the Blue Network, with flagship station WJZ (Newark), to candy manufacturer Edward Nobel—the Lifesaver candy entrepreneur.

THE AIRWAVES BECOME REGULATED

The Radio Act of 1927

Without some regulation of radio, it was impossible for the listener to receive a clear radio signal. Radio stations and radio amateurs could broadcast on whatever frequencies they chose and could increase their transmitting power at will.

Secretary of Commerce Herbert Hoover brought the various broadcasting interests together in Washington for a long series of meetings. After these meetings, the first comprehensive radio legislation was adopted by Congress—the Radio Act of 1927.

The new act established a five-member Radio Commission[6] and gave the

6. The Federal Radio Commission (FRC) was originally conceived as a one-year commission designed to operate with little financial and clerical support. The job was obviously too big for a temporary commission and the FRC became a permanent body on December 12, 1929. The FRC became the FCC in 1934.

commission power to regulate broadcasting and guarantee operations within the guidelines of the act. The Radio Act of 1927 also clearly established the principle that the public should benefit from this great force called radio and that all commercial radio operations should serve the needs and interests of the general public.

> The concept of the "public interest, convenience, and necessity" frequently appears in the act. Congress wanted no one to mistake the fact that the air waves were a public trust to be used for the public good. The provision of the act considering station licenses reads: "The commission, if the public convenience, interest, or necessity will be served . . . shall grant . . . a station license. . . ."[7]

The Communications Act of 1934

In 1934, Congress took another step in regulating broadcasting and passed the Communications Act of 1934. The act is founded on the principle that broadcasting is a privilege, and since the frequency spectrum is limited, those who are granted a license to broadcast should be more heavily regulated than are owners of other forms of media.

The Communications Act of 1934 established a seven-member Federal Communications Commission (FCC) charged with regulating broadcasting — licensing stations and broadcast operators, maintaining equipment standards and inspecting broadcast equipment, maintaining programming regulations, and, most importantly, guaranteeing that broadcasting in the United States serve the public interest.

Radio has been deregulated in recent years. The renewal application for radio stations has been significantly shortened; the length of the license period has been extended to seven years. A new communications act, or at least a rewrite of the 1934 act, appears to be in the offing. New technology — including cable television, fiber optics, and satellite communications — has dated the 1934 act. Any new broadcasting regulations will, however, receive extensive review both inside and outside the radio industry.

THE HEYDAY OF RADIO

Radio Broadcasting in the 1930s

The 1920s laid the groundwork for the expansion of radio. By the early 1930s, radio was a companion in nearly every household.

The nation was in the throes of a great depression with thousands of individuals jobless and many families without the essentials for daily survival. The

---

7. *Wireless World,* March 16, 1932, p. 283.

nation was nurtured in its national crisis by President Franklin Delano Roosevelt who, in 1933, began a series of what have been called "fireside chats" on radio. In Europe, Hitler and Mussolini were heard in public appearances with the roar of hysterical masses as the background. In sharp contrast, Roosevelt spoke in well-modulated, assured tones that gave the public confidence in itself, in its leaders, and in its nation. Roosevelt is said to be the first, and perhaps the best, master of political broadcasting. He is reported to have leaned into the microphone and chatted quietly with the American public as if he were a guest in the parlor. Radio is credited with being a vital force that held America together during the depression years.

During the 1930s, radio was king. Radio reigned over the American public, providing information, entertainment, and companionship. One listener who lived through radio's golden years wrote:

> The golden age of radio . . . spanned two decades. It began around 1929 with the arrival of Amos 'n' Andy. It ended around 1948 with the reign of Uncle Miltie on the tube. In tens of millions of American homes, the radio was the centerpiece. Whether in kitchen or front parlor, the Atwater Kent, the Philco console, or some other pridefully dusted set was *watched,* not just listened to. . . . Everyone had a favorite radio program, the one that just couldn't be missed.[8]

The 1930s were indeed the heyday of radio. Exhibit 1.7 shows some of America's favorite radio performers. By the close of the 1930s, Americans were thoroughly hooked on radio. One listener in its early years recalls the medium in these words:

> Unless one can remember network radio personally, he cannot grasp the place and power it once held in American life. . . . Radio drama . . . was theatre of the imagination. . . . Humor, too, was at a high level. . . . One of the most important components of radio's content was music. . . . It gave the U.S. a cornucopia of musical riches, from the Grand Old Opry to grand opera, from Woody Herman to Arturo Toscanini. And it was "live," not recorded music. . . . Network radio, although it was commercially sponsored, to a large extent lived up to the expectations of its pioneers. It was broadly educational—a cultural resource of inestimable value.[9]

Radio in the 1940s

During the 1940s, radio again served the nation in a vital capacity. Radio brought World War II into American parlors.

On December 8, 1941, President Roosevelt went before Congress to declare that the nation was in a state of war. Seventy-nine percent of the homes in

8.   Irving E. Fang, *Those Radio Commentators!* (Ames, Iowa: Iowa State University Press, 1977), p. 3.
9.   Gene Lees, "The Heyday of Network Radio," *High Fidelity,* December 1977, p. 19.

Exhibit 1.7. Radio's Heyday

Favorite programs on radio included music, drama, news, and commentaries. Some of the favorite performers included: Bing Crosby, Ed Wynn, Rudy Vallee, Freeman Gosden and Charles Correll as Amos 'n' Andy, Fred Allen, Jack Benny, and Bob Hope. Franklin Roosevelt commanded attention in American homes with his "fireside chats." The radio audience also paid loyalty to news personalities and commentators such as H. V. Kaltenborn and Edward R. Murrow. (*top*) Ned Weaver as "Dick Tracy"; (*bottom*) Ted Husing with Eddie Cantor. (CBS photos courtesy of Broadcast Pioneers Library.)

America tuned in to hear the President in somber tones declare war. As American men and women went abroad to do battle, the nation relied on radio for direct reports from the scenes of battle. By 1946, one survey indicated that 74 percent of all Americans listened to at least one hour of radio each day and 25 percent of all Americans listened to three or more hours of radio each day. At the time of the survey, 74 percent of Americans had not read a book in the last month, and 40 percent said that they did not regularly read any magazines.[10]

Americans learned the comfort of having an abiding source of information and entertainment. In 1946, one columnist expressed the sense of dependency that the American public had developed for radio during the war years and the sense of comfort that the new medium brought:

> There was practically no time of the day or night that one could not tune to some station and find out what was happening right up to the last minute.... Newspapers had and continue to have their functions.... But they could not satisfy the hunger for the immediate that was in all of us.... There was comfort in the very existence of radio, the knowledge that no matter where you were, in your car, or shipboard, traveling on a train, or visiting the home of a friend, you were never out of touch with events.[11]

Certainly not all programming during these years dealt with war issues. Much of it was pure entertainment and pure escapism. The networks were making a profit by providing Americans with news, information, and entertainment. In 1944, the gross billings for all radio networks amounted to nearly two hundred million dollars.[12]

Broadcasters were quick to assert the power of radio to reach the masses. In *Broadcasting* magazine, radio operators informed advertisers that for just one dollar, the national advertiser could reach 114 families through magazines, 122 families through newspapers, and 189 families through radio.[13]

By 1950, commercial network radio was only twenty-four years old. Yet, it had defined its role in American life and had sustained the nation in times of crisis—an economic depression, a war, national disasters, and other emergencies. The medium had demonstrated its power by its almost total saturation of the American market. By 1950, radio was faced with the crucial issue of redefining its role in American life. The nation was into a new broadcast era and the heyday of national radio was drawing to a close. Television was to be the new national medium and radio had to hunt for a new niche in the video society.

10. Paul F. Lazarsfeld and Patricia L. Kendall, *Radio Listening in America* (New York: Prentice-Hall, 1948), pp. 2–3.
11. *Broadcasting,* January 7, 1946, p. 26.
12. Charles A. Siepmann, *Radio's Second Chance* (Boston: Little, Brown, 1946), p. 66.
13. *Broadcasting,* January 14, 1946, p. 15.

RADIO IN THE VIDEO AGE

The Decline of the Radio Networks

During World War II, television had been waiting in the wings for its public debut, but the war efforts put television's public introduction off until the late 1940s. When the war ended, the number of applications for television stations increased rapidly, and radio found itself in a new era—the video age.

Many radio station owners were not worried by television's inroads, however. This was largely because many of the old radio station owners were owners of new television station licenses. If radio revenues were lessened by television, these radio/television station owners were preparing themselves for the loss of revenues from one investment and the gain from another. The radio owners who were in a dangerous position were those who did not own television stations.

There was no doubt that television was in the communications industry to stay and that it intended to win every advertising dollar it could. Calling television's growth "sensational," *Broadcasting* reported in early 1953 that television had grown from a fifty million dollar manufacturing industry in 1947 to a billion dollar industry.[14]

Even with television's rapid growth, radio revenues were not severely affected until the mid-1950s. The decrease in radio revenues was from the loss of national network advertising; local radio advertising sales were not severely adversely affected by television. Exhibit 1.8 indicates radio time sales from 1935 to 1965 and shows that radio's toughest years were 1954–1960.

A study of Exhibit 1.8 indicates that radio revenues were hurt most directly from the loss of national network radio advertising. Radio's biggest year for national sales was 1948, with the network advertising revenues at over 133 million dollars. This figure can be compared with radio's poorest year for network sales—1960—with total network advertising sales at a mere 35 million.

During the worst years of network radio losses (1954–1960), one individual observed in retrospect that network radio was held together more by "fear than by hope." During these hard years, the industry did not wish to publicize its losses, but when the broadcasters finally began to reflect on the tribulations they had endured during the last half of the 1950s, the tenuous status of network radio was exposed.

Television did not displace radio as the prime evening diversion in American homes overnight. The process took over a decade. But television was growing rapidly for a young upstart in the communication industry. In January 1952, *Broadcasting* reported that the number of advertisers using television had grown from 727 in January 1949 to 6,365 in January of 1952.[15]

14. *Broadcasting,* January 5, 1953, p. 34.
15. *Broadcasting,* January 7, 1952, p. 6.

Exhibit 1.8.
Radio Time Sales 1935–1965[1]

Year	National Network	% Change from Previous Year	Local	% Change from Previous Year	Total	% Change from Previous Year
1935	$ 39,737,867	—[2]	26,074,476	—	79,617,543	—
1936	—	—	—	—	—	—
1937	56,192,396	+ 41.4	35,745,394	+ 37.1	117,908,973	+ 48.1
1938	56,612,925	+ 0.7	32,657,349	− 8.7	117,379,459	− 0.6
1939	62,621,689	+ 10.6	37,315,774	+ 14.2	129,968,028	+ 10.7
1940	71,919,428	+ 13.1	44,756,792	+ 20.0	155,686,247	+ 20.5
1941	79,621,534	+ 10.7	51,697,651	+ 15.5	179,753,217	+ 15.4
1942	81,744,396	+ 2.7	53,898,916	+ 4.2	190,147,052	+ 5.8
1943	99,389,177	+ 21.6	64,104,309	+ 18.9	228,102,164	+ 20.0
1944	121,757,135	+ 22.5	84,960,347	+ 20.3	287,642,747	+ 26.1
1945	125,671,834	+ 3.2	99,814,042	+ 17.5	310,484,046	+ 7.9
1946	126,737,727	+ 0.8	116,380,301	+ 16.6	334,078,914	+ 7.6
1947	127,713,942	+ 0.8	147,778,814	+ 27.0	374,086,686	+ 12.0
1948	133,723,098	+ 4.5	170,908,165	+ 15.6	416,720,279	+ 11.4
1949	128,903,467	− 3.6	182,144,301	+ 6.5	425,357,133	+ 2.1
1950	124,633,089	− 3.3	203,210,834	+ 11.6	453,564,930	+ 6.6
1951	113,984,000	− 8.5	214,519,000	+ 5.6	456,543,000	+ 0.6
1952	102,528,000	− 10.0	239,631,000	+ 11.7	473,151,000	+ 3.6
1953	92,865,000	− 9.4	249,544,000	+ 4.1	477,206,000	+ 0.9
1954	78,917,000	− 15.0	247,478,000	− 0.8	451,330,000	− 5.4
1955	60,268,000	− 23.6	272,011,000	+ 9.9	456,481,000	+ 0.7
1956	44,839,000	− 25.6	297,822,000	+ 9.5	491,707,000	+ 7.7
1957	47,951,000	+ 6.9	316,493,000	+ 6.3	537,664,000	+ 9.3
1958	42,786,000	− 8.7	323,207,000	+ 2.0	541,664,000	+ 0.9
1959	35,633,000	− 23.4	359,138,000	+ 11.0	582,914,000	+ 7.6

20

Year						
1960	35,026,000	− 1.7	385,346,000	+ 7.3	622,474,000	+ 6.8
1961	35,837,000	+ 2.3	384,053,000	+ 0.3	617,242,000	− 0.9
1962	37,326,000	+ 4.2	419,468,000	+ 9.2	665,249,000	+ 7.8
1963	41,797,000	+12.0	449,717,000	+ 7.2	711,741,000	+ 7.0
1964	43,783,000	+ 4.8	487,947,000	+ 8.5	763,768,000	+ 7.3
1965	44,127,000	+ 0.8	549,663,000	+12.6	838,704,000	+ 9.8

[1] Source: The above figures were from FCC records. The above information was published in *Broadcasting*, February 10, 1969, p. 53.
[2] − indicates that the information was not available for that year.

NBC radio claimed its worst year in revenues was 1956, when losses rose to $3.4 million. Mutual Network radio announced losses of a million dollars or more each year for ten years. An ABC official revealed that in 1958, ABC had considered getting out of radio altogether, since its annual losses were running about four million dollars.

But the 1960s brought economic stability to the radio networks. By 1963, *Broadcasting* magazine was reporting, "Network radio . . . appeared to have a firm grip on what [a decade earlier] seemed completely out of reach—not only survival, but profitability."[16]

Radio Changes Formats

The early programs on television were similar to the competition on radio—comedy, variety programs, musicals, dramas, and news. When the audiences could see the programs, not just listen to them, radio was at a serious disadvantage. The medium of radio had to compete by offering a new product, and that is exactly what it did. It developed an almost totally local programming format. The disc jockey fit the bill nicely. The concept of the disc jockey, one who sat at a radio control board cueing records and chatting with the audience, was particularly attractive because the costs of this form of programming were low and local advertising revenues would pay for production.

The disc jockey has been uniquely associated with Top 40 radio, which developed in the 1950s. The beginning of Top 40 radio dates to 1949, when KOWH (Omaha) began to be selective in its popular music format. The true Top 40 playlist—the most popular records (based on record sales)—was defined by KOWH in 1956. The Top 40 sound involved what has been described as a "clock hour" formula.

> The Top 40 station's strict formula was built around a "clock hour" which called for certain events to occur at very carefully prescribed times. For example, a Top 40 station might specify that the song played at the "top" of the hour be a hit from the top 10, and that its rhythm be uptempo. The next record might also be a hit, or perhaps a "hitbound" or a familiar "oldie," but whatever the variation, each different type of music to be played would be prescribed. In addition, stations with "clock" formats scheduled times for non-musical features such as weather, traffic reports, sports scores, and even news bulletins.[17]

The Top 40 stations were not embarrassed to launch promotional campaigns that sent listeners on treasure hunts, or on hunts for disc jockeys traveling in mobile units, or that required listeners to place bumper stickers on their

16. *Broadcasting,* October 7, 1963, p. 33.
17. David T. MacFarland, "Up from Middle America: The Development of Top 40," in Lichty and Topping, *American Broadcasting,* pp. 401–402.

automobiles, or brightly colored balls or flags on the automobile radio aerial. Top 40 radio made no apologies for being competitive and trying to build a loyal audience. The Top 40 format saved the local radio station from economic harm during radio's transition period, 1955–1965.

Technology Defines a New Role for Radio

A technological development also served to increase radio's flexibility. The development was the transistor, introduced by Bell Laboratories in 1948. The utilization of the transistor increased during the 1950s as radio was redefining its role. Now, the radio station owners could compete for the audience at the beach, in the backyard, or almost anywhere. Transistor radios and automobile radios were extremely important to the success of Top 40, since Top 40 stations found that 50 percent of their audience was outside the home.[18]

The Target Audience Defines a New Role

The local radio broadcaster survived by segmenting the audience—deciding what a large segment of the potential audience wanted to hear and playing to that segment. The target audiences were sometimes defined by time of day, with the broadcasters appealing to the labor force in the morning hours as they dressed and drove to work, appealing to women eighteen to forty years of age during the day, and reserving the evening hours for teenagers who did homework with the aid of the transistor radio. Whatever the format, the day of the national radio audience was dead and the era of the local target audience had arrived.

Although Top 40 and other limited formats saved local radio, these formats have not been without their critics. One has called Top 40 "censored radio": "By the fiat of a station's owner, everything except that predetermined number of songs that are the most commercially successful at the moment is, in effect, banned from the air."[19] Indeed, Top 40 formats have, to some extent, limited the scope of American musical exposure.

One format that has proven to be very profitable for stations and less restrictive than Top 40 is the middle-of-the-road format (MOR). The MOR sound has been described as "standard vocal and instrumental arrangements of familiar music," and as "listenable," and as "familiar sound that never irritates."[20] By the mid-1960s, MOR radio had become one of America's great success stories. The format was making station owners wealthy and was pleasing radio audiences. While the programming formats of many stations are said

18. *Ibid.,* p. 403.
19. Gene Lees, "The Dotage of American Radio," *High Fidelity,* January 1978, p. 29.
20. *Broadcasting,* October 23, 1967. See the special report on MOR radio.

to be "vertical"—programming all jazz, all news, all rock, or all classical—MOR station formats include some rock, some jazz, some country, some feature stories, and some news.

With new formats, radio survived the inroads made by TV. A recent survey indicates that nearly every American home has several radio receivers and that radio's average daily audience is the biggest since the advent of television. Individuals twelve years of age and older listen to radio an average of 3.08 hours each day; persons eighteen to twenty-four years of age listen to radio an average of 3.54 hours each day, and radio listening for this group is expected to increase to 5.02 hours a day during the next decade.[21] Appealing to selected target audiences saved the radio industry and continues to increase revenues for radio.

NPR Defines a New Role for Radio

One exception to the commercial formats and the push for advertising dollars is National Public Radio (NPR). Created in 1967, public broadcasting offers Americans concerts by great orchestras, dramas, and informational programs like the award-winning "All Things Considered." Public broadcasting stations do not solicit advertising dollars; they survive on federal and state grants, grants from business and industry, and contributions from local citizens. Stations affiliated with NPR are frequently called educational stations, and their number has grown to nearly one thousand. NPR stations offer local listeners an alternative to all rock, all jazz, all news, or MOR formats.

RADIO TODAY AND TOMORROW

Technology Introduces Change

Today, all indicators predict continued good health for radio in the United States. A big story in radio today is the technological changes that are altering the day-to-day operations at the local station. One major change is that many stations are moving to automation.

An automated station is one that has all or part of its operation controlled by computer. The computer is programmed to play a particular tape at an exact moment or to load a cartridge for play at a specific time. A visitor standing in the computer control center of an automated station will see tapes spinning and cartridges loading and unloading, all without the interference of human hands. The human input came when the station operator programmed the computer with the information to select the desired program segments. At some sta-

21.   National Association of Broadcasters, *Radio in 1985,* Washington, D.C.: National Association of Broadcasters, 1977.

tions, not only does the computer assist in programming, but it controls station operations from the writing of a commercial order, to airing the commercial, to issuance of the bill and, finally, to recording the entry into the station log book.

Automation is reaching into the small markets as well as into the large markets. According to one industry publication, the stations are installing the automated systems for a variety of reasons: "smoother, better-controlled productions," "a big-city sound in a rural community," "weekend programming without hiring extra help," or "to keep the personnel lists from growing out of hand."[22] Whether or not computerization will improve working conditions in the radio industry, automation is here to stay.

Technological changes appear to be the major outlook for radio in the next few decades. In a report issued by the National Association of Broadcasters, radio industry leaders predicted:

> Quadraphonics will be in general use in FM radio.
>
> Stereo will be in general use by AM stations.
>
> All radio stations in a given market may be carried by a cable system. This would introduce new competitive factors since the number of available stations would increase. The radio audience may be even further segmented.
>
> Many radio stations will receive signals directly from satellite. Audio quality will increase since a higher quality signal can be sent via satellite than via hard-line or terrestrial transmission.[23]

New Options for New Audiences

Not only will technological advances continue, but it is predicted that the radio industry as a whole will experience growth. In the 1960s there were approximately 156 million radio receivers in working condition in the United States; in the 1980s that number is projected to reach 560 million or 2.4 radio receivers for each person in the United States. Radio audiences are projected to increase to coincide with the increased number of receivers.

The radio listener will have a greater diversity of programming than in past years. Because radio has again proven itself to be profitable, investors are putting money into new programming ideas. The next decade will bring new radio networks and new radio programming companies. These new programming options will indeed be another big story in the progression of radio.[24]

In 1897, when Guglielmo Marconi took out a patent for his wireless apparatus, he did not envision the kind of radio industry that we have today in the

22. "Radio Automation Gives Extra Arms to Talented People," *Broadcast Management/Engineering,* July 1976, p. 56.
23. National Association of Broadcasters, *Radio in 1985.*
24. See the special section on "Radio, 1981" in *Broadcasting,* August 17, 1981. See also the "Riding Gain" department in each 1983 issue of *Broadcasting.*

United States. Radio has indeed become a vital force in American life. Americans depend on radio for fast-breaking news, for weather information, for information on business and industry within the local community, for entertainment, for companionship. The changes that are taking place in the radio industry will allow radio to continue to evolve to meet the challenges of new technology and to more fully serve the unique needs of the audiences.

BIBLIOGRAPHY

Archer, Gleason L. *History of Radio to 1926*. New York: American Historical Society, 1938. (Reprinted by Arno Press, 1971.)

Barnouw, Erik. *A Tower in Babel: A History of Broadcasting in the United States*. vol. 1, to 1933. New York: Oxford University Press, 1966.

Barnouw, Erik. *The Golden Web: A History of Broadcasting in the United States,* vol. 2, 1933–1953. New York: Oxford University Press, 1968.

Blake, George Gascoigne. *History of Radio Telegraphy and Telephone*. London: Chapman and Hall, 1928. (Reprinted by Arno Press, 1974.)

Broadcasting Magazine, 1946–1983.

Carneal, Georgette. *A Conqueror of Space: A Bibliography of the Life and Work of Lee de Forest.* New York: Horace Liverright, 1930.

Davis, Harry P. "The Early History of Broadcasting in the United States." In *The Radio Industry,* Chicago: Shaw, 1928, pp. 189–222. (Reprinted by Arno Press 1974.)

de Forest, Lee. *Father of Radio: The Autobiography of Lee de Forest.* Chicago: Wilcox & Follett, 1950.

Donaldson, Francis. *The Marconi Scandal.* New York: Harcourt, 1962.

Dreher, Carl. *Sarnoff: An American Success.* New York: Quadrangle/New York Times Book Co., 1977.

Dunlap, Orrin E., Jr. *Dunlap's Radio & Television Almanac.* New York: Harper, 1951.

Fang, Irving E. *Those Radio Commentators!* Ames, Iowa: Iowa State University Press, 1977.

Fessenden, Helen M. *Builder of Tomorrows.* New York: Coward-McCann, 1940. (Reprinted by Arno Press, 1974.)

Geller, Harvey. "Nathan Stubblefield: The Radio Prophet of the Kentucky Fields." *High Fidelity,* November 1977, pp. 79–83.

Harlow, Alvin F. *Old Wires and New Waves.* New York: Appleton-Century, 1936. (Reprinted by Arno Press, 1971.)

Head, Sydney W., with Sterling, Christopher. *Broadcasting in America: A Survey of Television, Radio, and New Technology,* 4th ed. Boston: Houghton Mifflin, 1982.

Hilliard, Robert L. *Radio Broadcasting.* 2d ed. New York: Hastings House, 1974.

Jackson, Joseph H. "Should Radio Be Used for Advertising?" *Radio Broadcast,* November 1922, pp. 72–76.

Lazarsfeld, Paul F., and Field, Harry N. *The People Look at Radio.* Chapel Hill: University of North Carolina Press, 1946.

Lazarsfeld, Paul F., and Kendall, Patricia L. *Radio Listening in America.* New York: Prentice-Hall, 1948.

Le Duc, Don R., and McCain, Thomas A. "The Federal Radio Commission in Federal Court: Origins of Broadcast Regulatory Doctrines." *Journal of Broadcasting* 14 (Fall 1970).

Lees, Gene. "The Dotage of American Radio." *High Fidelity,* January 1978.

Lees, Gene. "The Heyday of Network Radio." *High Fidelity,* December 1977.

Lichty, Lawrence W., and Topping, Malachi, eds. *American Broadcasting.* New York: Hastings House, 1975.

Lyons, Eugene. *David Sarnoff.* New York: Harper, 1966.

MacLaurin, W. Rupert, with Harmon, Joyce R. *Invention and Innovation in the Radio Industry.* New York: Macmillan, 1949. (Reprinted by Arno Press, 1971.)

Marconi, Degna. *My Father Marconi.* New York: McGraw-Hill, 1962.

"One Newscaster Plus One Computer: Flexible Error-Free Operations for Mutual." *Broadcast Management/Engineering,* March 1977.

Raby, Osmond. *Radio's First Voice: The Story of Reginald Fessenden.* Macmillan of Canada, 1970.

"Radio Automation Gives Extra Arms to Talented People." *Broadcast Management/ Engineering,* July 1976.

Radio Broadcast, May 1922–April 1930. Garden City, N.Y.: Doubleday.

"Radio Currents." *Radio Broadcast,* May 1922, pp. 1–4.

Siepmann, Charles A. *Radio's Second Chance.* Boston: Little, Brown, 1946.

Wireless Age, 1918.

Wireless World, 1932.

QUESTIONS FOR REVIEW AND DISCUSSION

1. How did Marconi contribute to the development of radio?
2. How did the following individuals contribute to radio development in the United States?
   Lee de Forest
   Reginald Fessenden
   David Sarnoff
   Dr. Frank Conrad
3. How did the concept of radio change from 1896 to 1926?
4. How did radio amateurs influence the development of radio?
5. Why are these dates significant in the development of radio?
   1897
   November 2, 1920
   November 15, 1926
6. How was radio broadcasting accepted by the American public?
7. How were the ABC and Mutual networks formed?

8. What do you consider to be the most significant events in the development of radio broadcasting in the United States? Why?

9. What is the present legislation that governs broadcasting in the United States?

10. How would you characterize radio broadcasting in the United States in the 1930s?

11. How would you characterize radio broadcasting in the United States in the 1940s?

12. What were the worst years for national radio networks?

13. When did network radio begin to recoup from the economic losses inflicted by television?

14. When and why did Top 40 develop?

15. How did the transistor benefit radio station owners?

16. What is one criticism of Top 40 radio?

17. What is meant by an MOR format?

18. What are the technological changes predicted for radio in the next twenty years?

19. In general, what are future projections for radio broadcasting in the United States?

Chapter Two

THE SCIENCE OF RADIO

Radio is provided with its cloak of invisibility, like any other medium. It comes to us ostensibly with person-to-person directness that is private and intimate, while in more urgent fact, it is really a subliminal echo chamber of magical power....

Marshall McLuhan,
"Radio: The Tribal Drum"
Understanding Media: The Extension of Man, 1964

Radio, as we know it today, is far more complex and serves us in many more ways than ever envisaged by Marconi in his early experiments. While none of the basic ideas of radio waves has changed much from that time, our understanding of these ideas has changed. Advances in the uses of radio waves have gone hand in hand with the development of improvements within the broadcasting industry. This chapter discusses the physical principles of radio, an understanding of which will lead to an understanding of modern radio technology.

THE LANGUAGE AND MEANING OF RADIO WAVES

The Essence of Broadcasting

A radio broadcast is actually the transmission of radio waves. Radio waves are also called *electromagnetic waves* or *electromagnetic energy.* The components of electromagnetic waves are obviously the forces of electricity and magnetism.

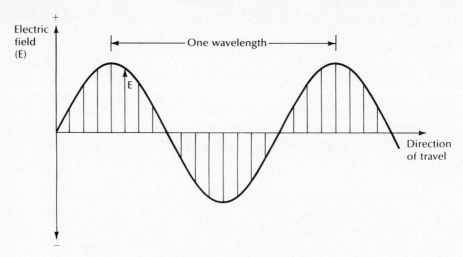

Exhibit 2.1. A Radio Wave

Exhibit 2.1 shows what an electromagnetic oscillation (radio wave) looks like. The electric field oscillates from positive to negative and back to positive again to form one complete oscillation. As Exhibit 2.1 also indicates, one complete oscillation or cycle is one radio wavelength.

Frequency and Wavelength

Two physical properties associated with radio waves are *frequency* and *wavelength.* The frequency of a wave is its number of oscillations or cycles per second. The wavelength, as indicated in Exhibit 2.1, is the distance between two consecutive maxima or "tops" of the waves.

Wavelength and frequency are related. The simple formula below shows this relationship:

$$V = W \times F$$
Where *V* is the velocity of the electromagnetic wave or the speed of light (186,300 miles per second),
W is the wavelength, and
F is the frequency.

This formula indicates that, for radio waves, the product of the wavelength and the frequency is equal to the speed of light. Wavelengths are expressed in meters,[1] or feet, and frequencies are expressed in cycles per second. Therefore,

1.   See Appendix A for a discussion of the metric system and conversion units.

the product of a radio wavelength times a frequency is some number of meters per second or feet per second, both of which are acceptable units of velocity.

Units for expressing frequency are hertz. A hertz is one oscillation or one cycle per second. Hertz is abbreviated as Hz. A larger unit for expressing frequency is the kiloHertz (KHz), which is one thousand oscillations per second. A megaHertz (MHz) is a million cycles per second, and a gigaHertz (GHz) is one billion cycles per second. This unit of measurement is named after Heinrich Hertz, who proved the existence of radio waves in 1887.

From the above discussion and the above formula, it becomes clear that if one knows the wavelength of a radio wave, the frequency can be obtained by dividing the speed of light by the wavelength; if the frequency is known, the wavelength is similarly obtained. Frequency and wavelength are dependent on one another and are equivalent ways of describing an electromagnetic wave.

Amplitude

There is an additional property of radio waves, called *amplitude*. The amplitude is the "size" or magnitude of the electric field and is a measure of the energy transported by the wave. The electrical power (number of watts) of a radio transmitter determines the amplitude or strength of the radio wave. A high power transmitter produces a very large amplitude radio wave. A low power transmitter produces a small amplitude radio wave. The distance over which a radio wave can be received is partially determined by the amplitude of the radio wave or the power of the radio transmission.[2] Exhibit 2.2 shows a large amplitude and a small amplitude wave. (Amplitude is sometimes expressed in dB–decibels. Appendix B provides a discussion of decibels.)

Three Characteristics of Radio Waves

At this point, we now understand a great deal about radio waves (or electromagnetic waves). 1) Frequency is a time characteristic of radio waves and indicates the number of times per second that a wave oscillates. 2) Wavelength is a spatial characteristic of radio waves and is expressed in feet or meters. 3) Every radio wave has a strength or amplitude. High power transmitters generate large amplitude radio waves while low power transmitters generate weak radio waves. Radio reception is partly determined by the amplitude of the radio wave.

Understanding these characteristics of electromagnetic waves is fundamental to understanding the production of radio waves. These concepts are at the heart of radio wave production and represent the fundamental science associated with radio transmissions.

---

2. The power of the transmitter or the energy transported in a radio wave is proportional to the square of the amplitude of the wave produced by it.

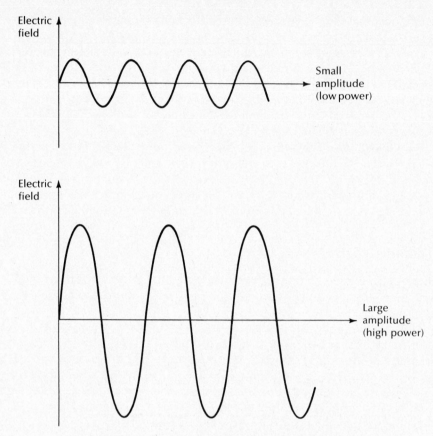

Exhibit 2.2.   Large and Small Amplitude Radio Waves

THE RADIO BANDS

The AM Band

In the radio-television portion of the frequency spectrum (Exhibit 2.3), there are two primary radio broadcast bands. The AM band is between 540 KHz and 1600 KHz. The FM band is between 88 and 108 MHz.

Within the AM band, the channels are spaced every 10 KHz apart, but the Federal Communications Commission (FCC) attempts to leave at least one unused channel between stations that may interfere with each other. There are 107 possible frequency allocations in the AM band.

Channels within the AM band have four possible classifications. A class one station is called a *clear channel station* and operates on 50,000 watts of power. These stations are designed to render service to very large areas, and there should be no interference from any other station. Their coverage sometimes ex-

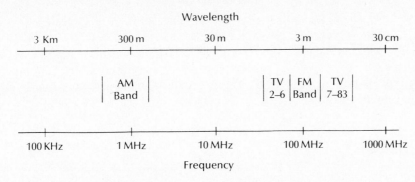

Exhibit 2.3. The Radio Bands

ceeds seven hundred miles in the evening hours. The United States has forty-five clear channel allocations.

A class two station is a *secondary station* that operates on 250 to 50,000 watts and may be subject to interference from class one stations. Class two stations are designed to render service over large areas. The United States has twenty-nine class two allocations.

A class three station is a regional station that operates with 500–5,000 watts of power. A class three station is designed to render service primarily to principal centers of population and to the suburbs that surround the populous areas. There are forty-one class three channels, housing more than two thousand stations.

A class four station is a local station. Class four stations operate on a maximum of 1,000 watts of power during the daytime hours and as low as 250 watts at night. These stations are designed to render service primarily to a city or town and to its immediate suburbs. Class four has the smallest amount of AM frequency space (six channels), but approximately 25 percent of all standard broadcast AM stations in the United States are licensed in this category. Each class four channel houses 150 or more local stations. Since the stations have a very limited range, many stations across the country can operate on the same frequencies.

The classes of stations have additional subdivisions, which further specify conditions of operations such as daily operating hours, specific power permitted, and other specifications designed to insure maximum radio service for each community with a minimum of interference between stations.

To operate over four thousand AM stations without interference, the FCC has taken steps to maximize the use of the frequencies.

1. Since AM signals are sometimes carried at great distances in the evening due to skywave activity in the ionosphere, almost half of the nation's AM stations are licensed to operate only during the daylight hours.

2. Limitations in transmitter power are intended to keep station transmitters within assigned frequency boundaries.

3. Directional antennas are used to control transmissions in the proper direction.

With these kinds of regulations, the AM band can serve local communities around the country.

The FM Band

The FM band is between 88 and 108 MHz, with channels currently assigned every 200 KHz, leading to a total of 100 channels. There are currently approximately 3,600 radio stations that broadcast on these frequencies, but the FCC has considered several proposals that would expand that number. If the proposals are introduced, frequency space would be available for an additional 500 to 1,000 new FM stations.[3]

Within the FM band, space has been made available for educational and noncommercial broadcasting. These are the stations that are frequently licensed to educational institutions and generally carry programs produced by National Public Radio (NPR). (See Chapter Seven for a fuller discussion of FM station operations.)

THE RADIO TRANSMITTER

Modulation

To accomplish a radio broadcast requires, first, a sound within the broadcast station. This sound can be a person at a microphone, a record, or a tape. This sound must somehow be combined with the radio wave so that the sound can be broadcast to radio receivers.

The process of combining the sound signal with the radio wave is called *modulation*. The radio wave that the sound signal modulates is called the *carrier wave* (or just *carrier*), since it carries the sounds from the radio studio.

The carrier wave is produced by a transmitter, and the transmitter combines the sound wave with the radio wave, i.e., modulates the wave. Exhibit 2.4 shows the components of a radio transmitter and demonstrates the process of modulation.

As can be seen in Exhibit 2.4, the oscillator produces a high frequency carrier wave and maintains the precise frequency on which the broadcast station is licensed to operate. The sound signal and the carrier wave enter the modulator

---

3. See "FCC Opens Up FM Spectrum-wide," *Broadcasting,* May 30, 1983, p. 31.

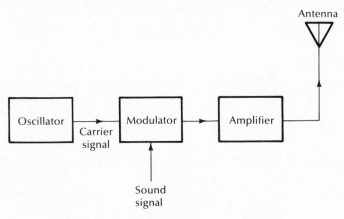

Exhibit 2.4. The Elements of a Transmitter

and are combined into one output that is then amplified and sent to an antenna where the modulated signal is radiated for pickup by the radio receiver.

Regulation and Broadcast Transmission

The entire modulation and transmission process is heavily regulated by the FCC to provide clear radio reception for the radio listener. Meeting these various regulations is largely the responsibility of the station engineer. If the regulations governing modulation and transmission are not met, the station may be severely fined by the FCC or may ultimately have its license revoked. Maintaining these engineering regulations is vital to station operations.

The amount of modulation of the sound signal is controlled by the operator at the audio board in the radio studio. The audio board operator must carefully monitor the program as it is being aired. If the sound signal is too large, the modulated signal will be distorted. A complete discussion of the audio board and its operations is given in Chapter 3. But monitoring the modulation process is the responsibility of the station engineer and the control board operator.

AM and FM Broadcasts

There are two kinds of modulation in use in radio broadcasting—*frequency* and *amplitude*. Frequency modulation is commonly known as FM, and amplitude modulation is commonly known as AM.

Frequency modulation is accomplished by varying the carrier frequency, whereas amplitude modulation involves varying the amplitude of the carrier wave. A pictorial representation of these two types of modulation is shown in Exhibit 2.5.

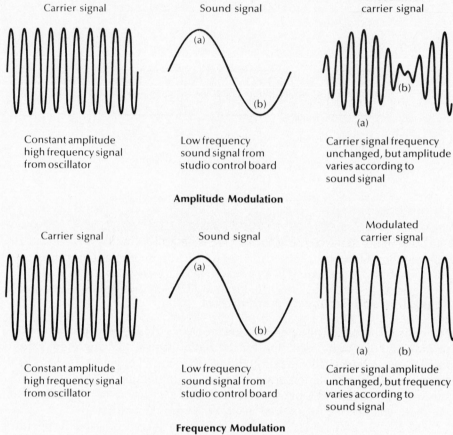

Carrier signal   Sound signal   Modulated carrier signal

Constant amplitude high frequency signal from oscillator

Low frequency sound signal from studio control board

Carrier signal frequency unchanged, but amplitude varies according to sound signal

Amplitude Modulation

Carrier signal   Sound signal   Modulated carrier signal

Constant amplitude high frequency signal from oscillator

Low frequency sound signal from studio control board

Carrier signal amplitude unchanged, but frequency varies according to sound signal

Frequency Modulation

Exhibit 2.5.   Types of Modulation

Notice in the exhibit that for amplitude modulation, region (a) represents a large amplitude signal and the amplitude of the carrier wave is large here; at (b) the sound signal has a large amplitude but is negative or "down," and the carrier wave after modulation has the smallest amplitude at this point. The carrier amplitude swings from large to small as the sound signal goes from positive to negative.

For frequency modulation, regions (a) and (b) represent the peaks in the change of the frequency of the carrier wave. (a) represents the maximum number of cycles per second, and (b) represents the minimum number of cycles per second.

If you have ever listened to AM radio during a thunderstorm, you are quite aware of the cracking and popping of the signal during lightning. The discharge of electricity that causes the lightning flash also produces radio waves with a very large amplitude. The lightning causes a quick upswing in the amplitude of

Amplitude: The strength of the radio wave. High power transmitters generate large amplitude radio waves while low power transmitters generate small amplitude radio waves.

Carrier frequency: The radio wave produced by the transmitter. The transmitter combines the carrier wave with the sound signal.

Current: The flow of electricity. The movement of positive and negative electric particles (electrons) that produces heat and magnetic fields and chemical transformations.

Frequency: The number of cycles per second of a radio wave. One cycle per second is called a hertz (Hz).

Modulation: The process of combining the carrier wave with the sound signal. There are two forms of modulation used in radio—amplitude modulation (AM) and frequency modulation (FM).

Power: The rate at which work is done or energy is expended. The basic unit of power is the *watt*.

Sound: The disturbance of particles in an elastic medium (such as air) that creates a wave-like pattern that vibrates and interacts with the ear. Sound can be characterized by wavelength, frequency, and amplitude.

Voltage: The electric potential or the amount of force pushing the electric current.

Wavelength: The horizontal size of a radio wave. Wavelength and frequency are inversely related.

Exhibit 2.6.   Radio Vocabulary

the radio signal. Since amplitude modulation involves variations in the amplitude, the AM radio pops and crackles as these large amplitudes or spikes of sound are received. Since FM does not rely on amplitude variations, it is less susceptible to interference in stormy weather.[4]

The stability of the FM signal is one of the major advantages that FM offers over AM. The AM signal can travel much farther than can the FM signal, however. While FM has some advantages over AM, AM has other advantages over FM. Exhibit 2.6 provides brief definitions of some technical terminology used in radio.

PROPAGATION OF RADIO WAVES

Three Modes of Travel

The travel of the radio wave from the transmitter to the radio receiver is called *propagation*. The method of transportation and the route of travel (propagation) of the radio wave are dependent upon the frequency. Different frequency waves have different characteristics and different frequency waves rely on dif-

4.   The quietness of the FM reception is also partly due to the presence of a limiter circuit in FM receivers that rejects very large amplitudes.

Exhibit 2.7.
Methods of Radio Wave Propagation

Type of Radio Wave	Frequency Range	Dominant Mode of Propagation	Common Usage
Very Low Frequency (VLF)	Less than 100 KHz	Groundwave	Navy
Low Frequency (LF)	100–600 KHz	Groundwave	Marine
Medium Frequency (MF)	600–3000 KHz	Groundwave, Skywave	Broadcast-AM
High Frequency (HF)	3–30 MHz	Skywave Directwave	Short Wave, Amateur, C.B.
Very High Frequency (VHF)	30–200 MHz	Directwave	VHF TV, FM, Air Band, Business, Fire
Ultra High Frequency (UHF)	Greater than 200 MHz	Directwave	UHF TV, Air, Business, Fire, Police

ferent methods of propagation. The three methods of travel for radio waves are via *groundwaves, skywaves,* or *directwaves.*

Exhibit 2.7 shows the common uses of the three methods of radio wave propagation, and shows the frequency ranges that correspond to the dominant modes of propagation. Exhibit 2.8 illustrates these three methods of propagation.

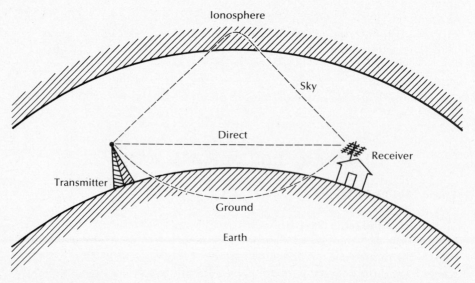

Exhibit 2.8.   A Visualization of Modes of Radio Wave Propagation

Groundwaves

Groundwaves are radio waves that travel through the earth or through water; low frequency waves travel most efficiently via groundwaves. One example of the use of groundwave propagation is the Navy's proposed Seafarer project. The Navy plans to bury thousands of miles of wire underground to form an antenna. Their goal is to broadcast in the VLF (very low frequency) range where groundwaves would be effective in reaching submarines anywhere in the world. Groundwave propagation is very effective for this type of communication. As Exhibit 2.7 indicates, groundwaves are useful in AM wave propagation but are not useful in FM wave propagation, since FM signals are in the high frequency ranges (88–108 MHz).

Skywaves

Skywaves are waves that are reflected from a layer of charged particles in the earth's upper atmosphere called the ionosphere. Skywaves are not always a reliable mode of propagation. Skywaves generate what are called *skipwaves,* which means that radio waves skip through the ionosphere and are sometimes propagated over very long distances but at other times are not propagated very far at all. If the ionosphere is heavily charged, skipwaves may propagate over thousands of miles and interfere with the reception of local stations.

The skipwave property of skywave propagation makes a distant AM station fade and flutter in response to changes in the earth's ionosphere. Another feature of skywave propagation is that the ionosphere is very responsive to activity on the sun. Streams of charged particles from the sun bombard the earth's atmosphere, which creates the ionosphere. In years of peak solar sunspot activity, skywaves (or skipwaves) may disrupt radio reception in portions of the spectrum not normally affected. Also, natural radio noises (pops and crackles in the AM bands) tend to become more pronounced during sunspot periods.

The skywave method of propagation is the most effective for receiving distant signals. It is not uncommon to receive AM or shortwave radio stations such as the BBC in London nearly anywhere in the world. But because of the skipwave characteristic of skywave propagation, a long distance signal may be intermittent. Also, because of the skipwave, a long distance signal may interfere with the reception of local stations.

Directwaves

Line-of-sight (or directwave) propagation means that the radio waves go directly in a straight line from the transmitter to the receiver. Directwaves are limited to approximately twenty-five miles between earth stations due to the curvature of the earth. (See Exhibit 2.8.) Because there is no dense matter in space (earth, buildings, and so on), directwaves can travel thousands of miles through space. Microwave frequencies used for satellite communications are in the multi-gigahertz range and are propagated by directwaves. New radio satellite networks rely on directwave propagation.

Directwave propagation is used for local radio stations—police bands, emergency bands, the television channels, and FM broadcasting. Since FM broadcasting relies on directwave or line-of-sight propagation, the FM signal is not propagated over great distances on earth. Because of the dependence on line-of-sight for propagation, FM stations are much less likely to interfere with each other than are AM stations, which are propagated by groundwaves and skywaves.

Since radio waves can interfere with each other, the antenna system for sending and receiving the waves becomes very important. It is possible to exercise some control over the propagation process in radio systems by careful design of the transmitter and the receiver antennas. Antenna function and placement are our next consideration.

ANTENNA CONSIDERATIONS

Antenna Height and Directionality

Two factors have large effects on radio wave propagation properties— transmitter antenna *height* and transmitter antenna *directionality*. For direct-wave propagation (FM and TV), antenna height is particularly important in determining the distance over which a radio signal will be received. For direct-waves, each additional hundred feet in antenna height will increase the range of the radio station by about four miles. For low frequencies of the AM band, antenna height is not important, since both groundwaves and skywaves are not significantly affected by antenna height.

The directionality of the transmitter antenna is also important in determining the coverage of a given radio station. Antennas can be made directional by using an array of radiators and reflectors spaced in such a way that the interference between the radiated and reflected waves results in cancellation of the waves along certain directions and reinforcement along others. This technique is sometimes used to prevent interference between two radio stations that have overlap in coverage area and are near each other in frequency allocation.

By focusing radio waves, very large antenna gain factors may be achieved. This gain is called *effective radiated power* or *E.R.P.* With proper antenna directionality, the E.R.P. can be much greater than the actual power delivered to the antenna by the transmitter. The FCC now regulates stations on the basis of their effective radiated power (E.R.P.), defined as:

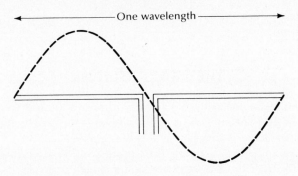

Exhibit 2.9.   Matching the Antenna to the Wavelength

E.R.P. = (actual power delivered to the antenna)
× (gain factor of the antenna)[5]

The gain factor can be as large as ten or more. Thus, a transmitter with the true power of 10 kilowatts (KW) could have an E.R.P. of 100 KW or more with a well-designed antenna system.

The above discussion involved height and directionality considerations for the transmitter antenna. At the receiver antenna, these same points can be made. By putting the receiver antenna high above the ground, it is possible to receive directwaves over greater distances, and by using a directional antenna, gain factors similar to those at the transmitter can be achieved. Furthermore, highly directional (so-called *beam*) antennas can have the advantage of minimizing interference from unwanted waves.

Wavelength Considerations

If our discussion of antennas is to be complete, the concept of "matching" the antenna to the wavelength of the radio must be mentioned. At the transmitter, the antenna must be tuned to a length that allows the electromagnetic waves to "fit" without some fraction of a wavelength left over.

Exhibit 2.9 shows an example of how an antenna can be matched to the wavelength of the radio wave. This example is just one possibility; other possible configurations exist. An antenna that has not been tuned will frequently

5. Gain factors and many other quantities in radio and sound measurement are frequently expressed in units of decibels (dB). See Appendix B for a discussion of dB units.

provide adequate reception, but a properly tuned antenna will generally improve reception.

THE RADIO RECEIVER

Having transmitted a radio carrier wave with a modulated sound signal out from the transmitter antenna, we must now reverse the process at the receiving end. Exhibit 2.10 shows a diagram of a radio receiver.

The radio receiver looks to be a little more complicated than the transmitter since the receiver must boost very small signals that are often only a few millionths of a volt (microvolts), and the receiver must be able to tune in stations broadcasting over a large range of frequencies.

As Exhibit 2.10 indicates, an R.F. or radio frequency amplifier is placed right at the input to increase the receiver signal. Tuning circuits are placed immediately following the R.F. amplifier to allow the receiver to be tuned to receive a particular frequency and reject all others.

The demodulator extracts the original sound (audio signal) from the carrier wave. The audio amplifier then increases the amplitude of the audio signal to a level appropriate to speakers or headphones.

Housed within most radio receiver cabinets are two separate receivers—one for AM and one for FM. On AM/FM radios, when the switch to select AM or FM is operated, the switch selects different R.F. amplifiers, tuning circuits, and demodulating circuits. About the only parts in common are the audio amplifiers and the speakers.

THE PROCESS OF RADIO: THREE STEPS

As Exhibit 2.11 indicates, radio is a three-step process—production, propagation, and reception. In step one, a signal is created (*production*) within the broadcast studio with the spoken voice, music, or some other sound. The signal then goes to the transmitter, which modulates it on the carrier wave and sends it to the antenna system of the radio station.

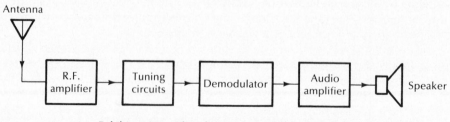

Exhibit 2.10.   The Elements of a Radio Receiver

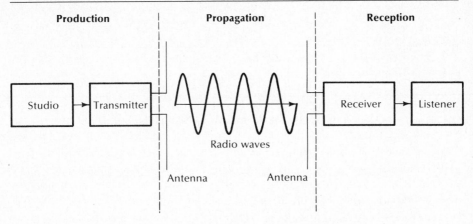

Production | **Propagation** | **Reception**

Studio → Transmitter

Radio waves

Antenna   Antenna

Receiver → Listener

Exhibit 2.11.   The Three Stages of Radio

At this point, the second step, called *propagation,* takes place. The accelerated electrons oscillating back and forth in the antenna produce a radio wave that propagates to a receiver antenna.

When the radio wave is received at the receiving antenna, the third step, *reception,* takes place. In the radio receiver, the radio signal is demodulated so that the sound signal is separated from the carrier wave and is amplified for the listener.

The mysteries of radio are less mysterious when one understands the basic science of radio. The science of radio transmission and reception must now be united with an understanding of radio broadcasting equipment. Chapter 3 discusses the equipment within the radio studio.

BIBLIOGRAPHY

Alten, Stanley R. *Audio in Media.* Belmont, Calif.: Wadsworth, 1981.

Adams, Arthur R.; Harper, Ralph; Johnson, Kenneth W.; and Walker, Willard. *Insights into Modern Communications: From Hi-Fi Sound to Laser Beams.* Minneapolis: Burgess, 1975.

Feldman, Leonard. *FM from Antenna to Audio.* Indianapolis: Sams, 1969.

Hasling, John. *Fundamentals of Radio Production.* New York: McGraw-Hill, 1980.

Johnson, Kenneth W., and Walker, Willard C. *The Science of Hi-Fidelity.* Dubuque, Iowa: Kendall/Hunt, 1977.

Marcus, Abraham, and Marcus, William. *Elements of Radio.* Englewood Cliffs, N.J.: Prentice-Hall, 1973.

Reyner, J. H., and Reyner, P. J. *Radio Communication.* London: Pitman, 1967.

Traylor, Joseph G. *Physics of Stereo/Quad Sound.* Ames, Iowa: Iowa State University Press, 1977.

Whetmore, Edward J. *The Magic Medium: An Introduction to Radio in America.* Belmont, Calif.: Wadsworth, 1981.

QUESTIONS FOR REVIEW AND DISCUSSION

1. Define the following: electromagnetic waves, frequency, wavelength, hertz (Hz).
2. Calculate the wavelength of one of your local AM stations. Repeat for an FM station.
3. Define amplitude. How are amplitude and power related?
4. List and briefly discuss three characteristics of radio waves that are outlined in this chapter.
5. What is the frequency range of the AM band? How many possible frequency allocations are there within the AM band?
6. What is the frequency range of the FM band?
7. Why is AM propagation more difficult to control than FM propagation? What steps has the FCC taken to control interference between AM stations?
8. What is meant by the term *modulation*?
9. What are the two kinds of modulation used in radio broadcasting? How do they differ?
10. What is meant by a clear channel?
11. Most AM stations in this country are in what classification?
12. What characteristics of AM reception make it more unstable than FM reception during a thunderstorm?
13. If a studio signal is overmodulated at the control board, what happens to the signal at the transmitter?
14. What are the characteristics of groundwaves, skywaves, and directwaves? What are some specific uses of these modes of radio wave propagation?
15. What two antenna factors have a large effect on FM radio wave propagation?
16. Define E.R.P.
17. What are the components of a radio receiver?
18. If you had to decide on the frequency allocation for a citizen's band two-way radio service, what would your choice be and why?
19. What are the three basic steps in the process of radio from the studio to the listener?

Chapter Three

RADIO
PRODUCTION ELEMENTS

*Radio production is both an art and a science. Proficiency in
the art of radio demands knowledge of the science of radio.*

Radio production work can be the most challenging and the most personally
satisfying of the electronic media because the equipment needed for radio production is relatively inexpensive and the creation of radio programs can be a
one-person operation. The least expensive form of radio production requires a
small tape recorder, a single-edged razor blade, and some editing tape. The
most sophisticated forms can utilize expensive stereo consoles (mixers),
automatic-cue cartridge machines, a series of turntables, stereo tape recorders,
a variety of microphones, frequency filters, and equalizers.

The amazing feature of radio production is that, with proper production
technique, radio programs produced on inexpensive equipment can be aired on
a standard broadcast radio station, whereas programs produced on very inexpensive video equipment generally cannot be aired on a standard broadcast
television station. This feature of radio production means that radio programs
can be produced almost anywhere with a minimum of equipment.

This chapter explores the various components of a radio station production
room. It examines the equipment that is needed for a shoestring operation and
the equipment that is likely to be found in large market radio. The chapter
begins with a discussion of the basic ideas of sound itself and then moves on to
examine the equipment designed to work with sound signals.

FUNDAMENTALS OF SOUND

Wavelength and Frequency

Sound, as far as we are concerned in radio production, is a pressure wave transported by the air around us. When we make a sound, the air is moved back and forth. When the movement is forward, the air is *compressed;* when the movement is away, the air is *rarefied*. Exhibit 3.1 indicates graphically what sound production looks like.

Since sound is a wave, it has a *wavelength* characterized by the distance in the air between two consecutive compressions or rarefactions. Just as in the case of electromagnetic waves, sound can be characterized by its *wavelength* and its *frequency*. Sound must occur in some medium, which is generally air. Sound will not propagate in a vacuum, since pressure fluctuations cannot be produced without something to compress. In air, sound has a speed of approximately 1,100 feet per second, which means it takes about five seconds to go one mile. Like electromagnetic waves, the product of the wavelength of a sound and the frequency of the sound always equals the speed of the sound.

Frequency × Wavelength = Speed (1,100 ft./sec. in air)

Sounds within the Range of Human Hearing

Within the range of human hearing, sound waves have wavelengths of about fifty feet at the low frequency end and about one-half inch at the high frequency end. The range of human hearing is approximately 16 Hz to 16,000 Hz. Most sounds that we listen to, such as music and human voices, have the majority of their sound energy concentrated between a few hundred and a few thousand Hertz.

If the sound source moves a great deal, it produces a large pressure fluctuation; this will be heard as a loud sound. If the pressure fluctuation is small, we hear it as a quiet sound. Thus, the amplitude of the sound wave is determined by the size of the pressure fluctuation. We hear large amplitude sound waves as loud sounds and small amplitude sound waves as quiet sounds.

Sound amplitudes are measured in units called decibels. Appendix B explains the decibel (dB) scale in some detail, but here it suffices to say that if the weakest sound we can hear is 0 dB, then the loudest sound we can hear without pain is 120 dB. Exhibit 3.2 shows some examples of various common sounds in the decibel scale.

Acoustics

Sound waves have an important property that must be remembered in sound recording and reproduction systems. When sound waves bounce off a surface

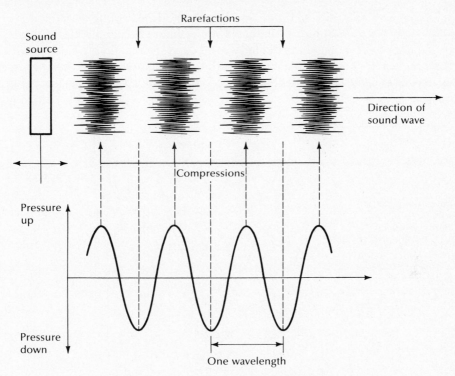

Exhibit 3.1.   Production of a Sound Wave

Exhibit 3.2.
Typical Sounds in Decibel Units

Intensity	Typical Sound
120 dB	Threshold of pain
110 dB	Jet plane taking off nearby
100 dB	Air hammer
90 dB	Outboard motor
80 dB	Loud music
70 dB	Street corner with traffic
60 dB	Conversation
50 dB	Busy office
40 dB	Average home noises
30 dB	Very soft music
20 dB	Quiet room
10 dB	Whisper
0 dB	Threshold of hearing

such as a wall, the reflected pressure fluctuation can interfere with the direct wave from the sound source. Exhibit 3.3 shows what can happen due to this interference effect. When the reflected wave's pressure maximum coincides with the directwave's pressure minimum, a partial cancellation of the sound occurs. On the other hand, if the reflected wave and the directwave have pressure max-

Interference leads to partial cancellation when the reflected wave has its compression point at the same place in the air that the direct wave's rarefaction occurs.

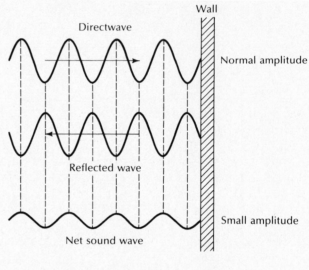

Interference leads to partial reinforcement when the reflected wave has its compression point at the same place in the air that the direct wave's compression occurs.

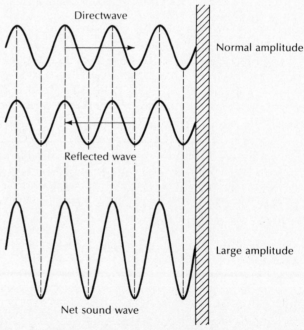

Exhibit 3.3.   Interference between Direct and Reflected Waves

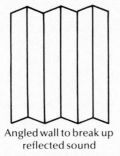

Angled wall to break up
reflected sound

Soft, sound-absorbing
material on wall to prevent
wave reflection

Exhibit 3.4.   Wall Surfaces that Reduce Sound Reflections

imums at the same point in the air, they will reinforce each other and the sound will be louder. This interference with the directwave can distort the original sound.

To avoid distortion and to reproduce the original sound, the sound studio will have material on the walls to reduce reflections of the sound. An irregularly shaped wall can prevent the reflected wave from reflecting back along the directwave. Exhibit 3.4 shows two types of wall surfaces that will reduce such sound reflections. There are many other interesting topics on the subject of acoustics; for more detail and discussion, see some of the references listed at the end of this chapter.

MICROPHONES

Microphones are essential instruments in sound recording. A microphone converts the mechanical pressure of sound waves into electrical energy, which is converted into electromagnetic waves to be picked up by the radio receiver. At the radio receiver, the electrical energy is converted back into the pressure waves that we hear as sound. The process of changing mechanical energy (fluctuating air pressure) into electrical energy, which is what the microphone does, is called *transduction.*

When selecting a microphone for a particular audio project, understanding some general characteristics of microphones will be useful. Microphones can be classified by: 1) the method of operation, 2) the response pattern, 3) unique response characteristics.

Method of Operation

A microphone operates in one of four ways — *magnetic induction, piezoelectric effect* (pronounced pē-ā-zō), *capacitance,* or *resistance.* The four methods of operation are quite different in physical detail, but all achieve the same goal of transforming sound energy into equivalent electrical energy.

The dynamic or magnetic microphones operate by the principle of magnetic induction. The motion of an electrical winding (a coil) within a magnetic field will *induce* a current or voltage difference within the winding. By attaching a diaphragm (a thin piece of metal) to a small coil and suspending the coil inside the magnet, an electrical signal can be produced that corresponds to the sound signal coming into the microphone.

In the dynamic microphone, the amplitude of the electrical signal is proportional to the amount of motion of the diaphragm. A large sound will produce a large movement in the diaphragm; a small sound will produce a small movement. The number of times per second that the coil moves in and out is determined by the number of sound waves per second arriving at the diaphragm — which is the frequency or pitch of the sound. The movement of the diaphragm should produce an electrical signal corresponding to the original sound in both amplitude and frequency.

The ribbon microphone is also a dynamic microphone with a somewhat different design. The ribbon microphone has a thin corrugated ribbon of a light alloy (sometimes this is aluminum) that moves in the field of a permanent magnet. The thin layer of alloy vibrates in response to the pressure difference between the two sides of the ribbon and produces an electrical current as it moves in the magnetic field. The dynamic and ribbon microphones have two commendable qualities: 1) they are rugged, and 2) they have good fidelity in their sound reproduction. Exhibit 3.5 demonstrates graphically how the dynamic or magnetic microphones function.

The crystal or ceramic microphone relies on a physical property called piezoelectricity whereby some materials when deformed will generate a

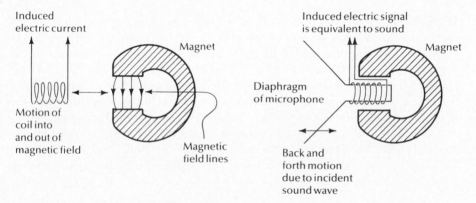

Exhibit 3.5.   The Dynamic or Moving Coil Microphone
The dynamic or moving coil microphone has a moving coil that moves in the field of a permanent magnet. The coil vibrates in response to the variations in the sound pressure. A ribbon microphone functions in a similar way to the moving coil microphone. A thin corrugated ribbon of light alloy moves in the field of a permanent magnet, also vibrating in response to the variations of the sound pressure.

voltage difference in proportion to how much they are deformed. By attaching a diaphragm to a small piece of piezoelectric material, the generated electrical signal will correspond to the motion of the diaphragm. Loud sounds move the diaphragm a large amount and thereby produce a large deformation of the piezoelectric material causing a large voltage. The number of times per second that the piezoelectric material is deformed determines the frequency of the voltage. Again, like the dynamic and ribbon microphones, the crystal and ceramic microphones *should* reproduce a sound equivalent in pitch and loudness to the sound from the source. Exhibit 3.6 demonstrates the functioning of the crystal and ceramic microphones. These microphones are extremely lightweight, portable, and inexpensive. They are frequently sold with cassette tape recorders. Unfortunately, the crystal and ceramic microphones do not always reproduce good sound quality and are not frequently used in radio broadcasting.

As shown in Exhibit 3.7, the condenser microphone depends on a basic physical property called *capacitance*—the ability to store energy. A capacitor is a device that can store electrical energy. A simple capacitor can be formed by holding two small plates next to each other (but not touching). The ability to store energy depends on the distance between the plates; if a small amount of charge is placed on one plate, it will induce a charge of the opposite sign on the other plate, and energy will be stored in the space between the plates. If one plate is moved closer to the other, the voltage difference will decrease because the capacitance is increased. On the other hand, if one plate is moved away from the other, the voltage difference will increase because the capacitance decreases. Again, this system transforms mechanical pressure into electrical

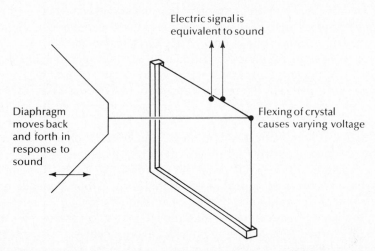

Exhibit 3.6.   The Crystal and Ceramic Microphone
In the crystal microphone, crystals or ceramic substances develop a voltage difference between the surfaces when subjected to mechanical stress.

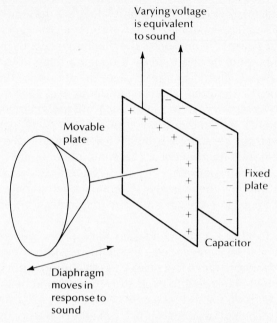

Varying voltage
is equivalent
to sound

Movable
plate

Fixed
plate

Capacitor

Diaphragm
moves in
response to
sound

Exhibit 3.7.   The Condenser Microphone
A condenser microphone consists of two closely spaced plates, one charged
positive and the other negative. One of the plates is free to move and is attached to
a diaphragm. When it moves in response to sound pressure, a varying voltage is produced between the plates.

energy. The number of times per second that the sound wave causes the plate
to move determines the frequency, and the larger the movement (louder the
sound), the larger is the voltage change across the capacitor.

The condenser microphones are frequently used for recording music. They
are relatively expensive and do not stand up to rugged treatment. Condenser
microphones are generally not recommended for field work since they may not
respond well to the jockeying they receive in transporting them or setting them
up for use.

The carbon microphone is the earliest form of microphone. It was introduced by Alexander Graham Bell and was perfected by Thomas Edison. This microphone operates on the simple idea that the resistance to the flow of current
through an element made up of small granules of carbon depends on how
tightly packed the granules are. If they are very loose and hardly touching one
another, there are few paths for the electrons to take to get through the carbon
granules. If, however, they are tightly packed and touching at many points, the
current can readily flow through the element. If a diaphragm is connected to an
element of carbon granules, the large sound waves will push the granules

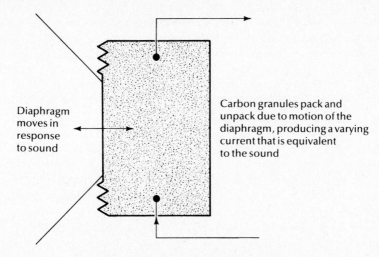

Diaphragm moves in response to sound

Carbon granules pack and unpack due to motion of the diaphragm, producing a varying current that is equivalent to the sound

Exhibit 3.8.   The Carbon Microphone
In the carbon microphone, granules of carbon pack together in response to motion of a diaphragm, causing a variation in the current flowing through the carbon element.

together tightly and produce a large change in the current flow, giving amplitude transformation properly. The number of times per second that the packing and unpacking occurs is determined by the frequency of the sound wave. The carbon microphone is shown in Exhibit 3.8. These microphones are no longer used in broadcasting, but they are used in telephones. Carbon microphones are far from ideal, since they are inclined to have a hiss and do not reproduce high frequencies very well. Exhibit 3.9 summarizes the various methods of operation of microphones.

Exhibit 3.9
Methods of Operation for Microphones

Physical Principle	Common Name	How It Works
Magnetic Induction	Dynamic, Ribbon	Moving coil or metal ribbon causes induced voltage
Piezoelectricity	Crystal, Ceramic	Deforming material causes voltage fluctuation
Capacitance	Condensor	Moving plate of capacitor causes voltage to change
Resistance	Carbon	Packing and unpacking of carbon granules changes resistance to current flow

The increasingly popular FM wireless microphones can use any of the above operational methods, although most are dynamic microphones. These microphones are wireless because the signal cable is replaced by a short-range radio link to the amplifier. Wireless microphones allow the performer much greater mobility than do cable microphones.

Response Patterns

Independent of the method of operation, all microphones transform the amplitude and frequency of a sound wave into electrical energy. How well they do this is partially determined by the response pattern and the response characteristics. The response pattern or *pickup pattern* is the area in which a microphone will accurately encode the sound. There are basically four types of microphone directional patterns: omnidirectional, bidirectional, unidirectional, and cardioid.

"Omni" comes from Latin and means all. Thus, the omnidirectional microphones pick up sound from all directions. Most dynamic and crystal microphones work in this way as do some condenser microphones.

Bidirectional microphones pick up sound from two directions. The directions are always opposite each other at 0 and 180 degrees. Bidirectional micro-

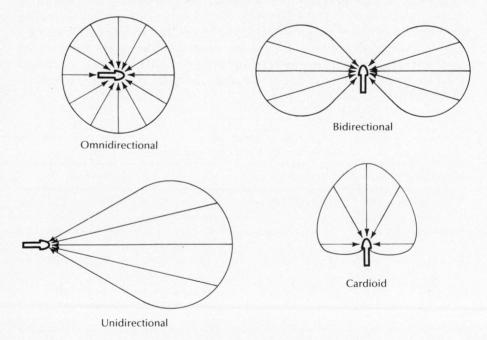

Omnidirectional

Bidirectional

Unidirectional

Cardioid

Exhibit 3.10.   Directional Patterns of Microphones

phones are sometimes called figure-eight microphones since a graph of the output looks like a flat figure eight. Ribbon microphones most often work in this manner.

Unidirectional microphones are highly directional. They pick up sound from directly in front of the microphone and are almost dead to sound from the sides and rear. These microphones are excellent for picking up one speaker in a crowd. A reporter may use a unidirectional microphone when trying to get the comments from a political leader on the floor of a national convention, for example. Ribbon and dynamic microphones can be unidirectional.

A cardioid microphone has a heart-shaped pickup pattern. The cardioid is frequently called a unidirectional microphone, but it has a greater coverage for sound pickup than do the highly directional microphones. The cardioid microphones pick up sound from the front and some sound from around both sides. Both condenser and dynamic microphones can have cardioid patterns. Exhibit 3.10 demonstrates the four types of pickup or response patterns.

Unique Response Characteristics

The other factors that are important in determining whether a given microphone is suitable for a particular job are frequency response, sensitivity, noise immunity, and distortion. Frequency response is an indication of the output voltage as a function of the sound frequency or pitch. The normal unit of sound level or voltage level is the decibel (see Appendix B). A decibel is a relative measurement with respect to some standard sound or voltage level. Exhibit 3.11 indicates what a frequency response diagram typically looks like.

In this exhibit, there is a region of frequency over which the curve is relatively "flat," around 0 dB; this is the acceptable portion of the microphone out-

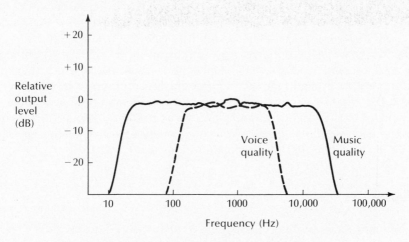

Exhibit 3.11. Frequency Response Diagram

put. At the ends of the "flat" response part of the curve, the microphone output drops rapidly to − 10 dB or worse. As explained in Appendix B, − 10 dB means that the signal is only one-tenth of what it is at 0 dB. Also shown in Exhibit 3.11 are two very different types of frequency responses: 1) music quality good response ("flat" response) from 50 Hz to 20,000 Hz (20 KHz) and 2) voice quality good response from about 200 Hz to 4,000 Hz.

Since a music quality microphone has the greatest frequency range, why not always use that type of microphone? There are numerous instances where having a wide frequency response actually interferes with the intended use of the microphone. To record an interview in a busy office, for example, it is best to have a microphone that responds to the range of human voice and rejects all other sounds, such as the hum of machinery. On the other hand, if you are recording a concert, you would not want to use a voice quality microphone since it would produce a tin can effect in the reproduced sound and would pick up only part of the musical sound range.

It is important to realize that no microphone is perfectly "flat" in the near 0 dB output range. A normal microphone will have a wiggly curve that stays around the 0 dB level and will only deviate significantly near the end of its acceptable frequency range. Also, when looking at manufacturer's specifications of frequency response, you may find that instead of having a scale centered on 0 dB, the output level will be in terms of the absolute voltage (referenced to a level of one volt) and the scale will range from about − 80 dB up to about − 50 dB. The same interpretation, however, still applies — namely, that the good response region is defined by nearly "flat" output across the frequency axis and that deviation of more than about 3 dB will have a noticeable effect in the reproduced sound.

The last point of discussion of unique response characteristics of microphones is the sensitivity of a microphone. A manufacturer's number indicates what absolute level of voltage the microphone will produce at the output for a standard sound level. Normal sensitivity ranges from − 70 dB to − 50 dB where the larger (less negative) number is more sensitive. These numbers are very small; this is why microphones are so sensitive to various kinds of noise generated internally and externally. Since these signals are so small, they must be amplified by a large amount to make them useful (up to a million times larger). This means that any unwanted signal will be amplified also and interfere with the desired signal. To avoid picking up unwanted signals such as hum or buzz, check regularly all cables and connections to be certain that the cable is in good shape and that the connections are solid.

Impedance

Microphones have another characteristic — *impedance,* the resistance to the flow of electricity in a cable. Microphones are high impedance or low impedance. *High impedance* microphones have greater resistance to the flow of electricity and, when used with long cable (fifteen feet or greater), they generally provide poorer quality sound than do low impedance mikes. *Low impedance* mikes are generally used for professional recording purposes, and high impedance mikes are used for home recording.

Impedance is measured in ohms. A low impedance mike is usually characterized by 150 ohms and a high impedance mike by 600 ohms.

A low impedance microphone should be used with a low impedance tape recorder; a high impedance microphone should be used with a high impedance tape recorder. A high/low mismatch will generally result in the loss of sound signal and sound fidelity.

Selecting a Microphone

In selecting a microphone, several questions must be asked and answered to get the best equipment for the job:

1. What is the exact nature of the job to be done?
2. Is excellent fidelity required?
3. Is cost a crucial consideration?
4. Is the ruggedness of the microphone a crucial consideration?
5. Must the microphone have unique features to accomplish the task adequately?

Answering such questions dictates the microphone to be used. If the microphone is to be used for recording music, a dynamic microphone or a condenser microphone will be the best choice since these microphones have excellent frequency response. If low cost is a major consideration, the crystal or ceramic microphones are the least expensive and provide good service for home, field recording, or nonbroadcasting purposes. If a rugged microphone is required, a dynamic microphone may be the best choice. If a microphone is used for a two-party interview, a bidirectional ribbon microphone would be an excellent choice. A microphone that will be used to pick up comments from an individual in a crowd should be a highly directional dynamic mike. The unique requirements of the situation and the environment must be the major considerations when selecting a microphone.

Microphone Placement

After the appropriate microphone has been selected, positioning it for the best possible sound pickup is the next consideration. Placing the microphone in the best position in relation to the sound source and the surroundings is called *microphone balance*. There are at least two objectives in microphone balancing: 1) to get the best possible sound with the least possible distortion, 2) to eliminate unwanted noises.

In his book *The Use of Microphones,* Alec Nisbett writes that "good balance can be judged only subjectively, and the best way of achieving it is by making direct comparative tests in good listening conditions between two (or more) microphones and moving one at a time until the best sound is achieved."[1] Nisbett suggests that in the listening comparison, the audio operator should test the following possibilities for improving the sound quality:

1. The direction from the source
2. Height and lateral position
3. The angle of a directional microphone
4. Directional characteristics of several microphones
5. The frequency response of different microphones
6. The use of additional microphones

Obviously, microphone balance is an art and a science that is mastered only by working with a variety of microphones in a variety of situations. Before creativity can have full rein in radio production, the uses of the equipment must be fully mastered, and microphones are no exception.

AMPLIFIERS: BOOSTING THE SIGNAL

The amplifier is essential to the production of the modulated signal. The small signal sources in the studio (microphones, turntable cartridges, magnetic tape playback heads) must all be amplified. The control board has several amplifiers built into it, and when you select various sources in the studio, what you are actually doing is switching inputs to the control board amplifiers.

As was just discussed for microphones, amplifiers must increase the amplitude of audio signals by as much as a million times. This high gain, as it is called, is usually obtained by a series of amplifiers. For example, if one amplifier produces a gain of 100, then two of them hooked up one into the other will produce a gain of 100 × 100 or 10,000, and three give a gain of 1,000,000. In state-of-the-art amplifiers, it is possible to obtain gains as high as one million using one small integrated circuit device called an operational amplifier (OP-AMP).

In some studios, you will find that the amplifiers are the older tube type that must be turned on early and allowed to warm up. These amplifiers generate a good deal of heat in the filaments of the tubes, and they generally require cooling fans and occasional tube replacements, and they fail more frequently due to component heat damage. Newer amplifiers, using transistors or integrated circuits (see Exhibit 3.12), are generally more reliable, more compact, and require less power to operate. Nearly all portable equipment uses transistor (solid

1. Alec Nisbett, *The Use of Microphones* (New York: Hastings House, 1974), p. 67.

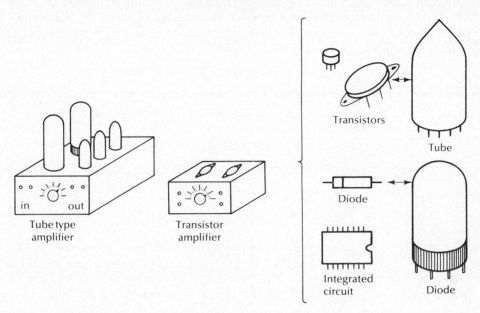

Exhibit 3.12.   Amplifiers: Comparisons of Transistors and Tubes

state) technology, since compactness and low power requirements are essential. Amplifier failure is usually indicated by: 1) no sound output, 2) hissing/crackling noises, 3) sporadic output.

RECORDING SYSTEMS

The Record and Playback Process

The basic component of any radio operation is the tape recorder. Without any other studio equipment, an individual can create a program for radio broadcast. A tape recorder can be taken to any location; the raw stock (magnetic tape) is inexpensive and reusable. The tape itself is easy to edit.

What happens at the recorder when the recorder is in the record mode? The audio signal is a fluctuating voltage or current. If this fluctuating current is put through a coil of wire, a changing current in the coil produces a changing magnetic field around it. If a piece of magnetizable tape is pulled past the region where this magnetic field is being produced, we can imprint a magnetic pattern upon the tape with this field. Exhibit 3.13 shows this process.

The playback is just the reverse of the record process. The moving tape with its regions of varying magnetic field induces changing currents in the coil in the playback head, which can be amplified and thereby become a useful audio signal. Exhibit 3.14 shows this process.

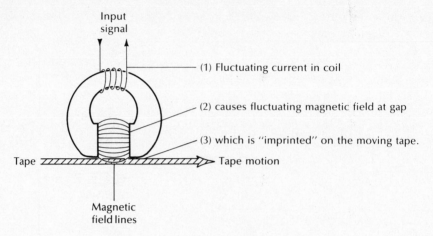

Input
signal

(1) Fluctuating current in coil

(2) causes fluctuating magnetic field at gap

(3) which is "imprinted" on the moving tape.

Tape   Tape motion

Magnetic
field lines

Exhibit 3.13.   Recording a Signal on Tape

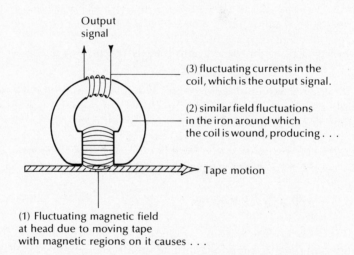

Output
signal

(3) fluctuating currents in the
coil, which is the output signal.

(2) similar field fluctuations
in the iron around which
the coil is wound, producing . . .

Tape motion

(1) Fluctuating magnetic field
at head due to moving tape
with magnetic regions on it causes . . .

Exhibit 3.14.   Playback of a Recorded Signal

The Tape Recorder Heads

A tape head is a shiny metal surface through which the electric signal is recorded on the tape in the record mode, and the electric signal is "read" by the tape recorder in the playback mode. Tape recorders have three heads—an erase head, a record head, and a playback head.

The erase head (degausser) is the first head over which the tape passes. In the record mode, the erase head sends out a high frequency signal that produces a rapidly varying magnetic field that "jumbles" the recorded signal on the tape. Any "order" from a previous recording is replaced by randomly arranged magnetic particles ready to be used in a new recording.

On every tape recorder the head arrangement is the same — the erase head, followed by the record head, with the playback head being the last head the tape passes over before moving to the take-up reel. Some recorders may have a combined record/playback head. In this case, the erase head would be the first head the tape passes, followed by the record/playback head.

Every tape head has a small gap where the magnetic field exists. The size of the gap determines the maximum frequency the recorder can handle. A buyer should look at the manufacturer's specifications to determine the cutoff for high frequencies.

Magnetic Tape

A piece of magnetic tape is shown in detail in Exhibit 3.15. As seen from the side edge of the tape, there are two layers: 1) a base material of mylar or polyester that determines the strength of the tape and 2) a thin layer of magnetizable material, usually an oxide of iron or a magnetic alloy.

Magnetic tape has magnetizable regions called *magnetic domains.* As the magnetic field of the head changes directions, so do the magnetic directions of the domains. In a very high frequency signal, the magnetic field of the head may change so fast that no new domains have moved into the head gap during the change, in which case the information is lost. If the tape recorder is set at a higher speed, more domains per second move across the record head and can record a higher rate of field direction changes. The more magnetizable domains per second that move past the record head, the higher the frequency audio signal that can be recorded.

A magnetic material with more magnetic domains will improve the high frequency response. There are several newer kinds of audio tapes that are composed of smaller domains and that have many more domains than ordinary recording tape. These tapes are frequently excellent for recording music.

Recording tape comes in a variety of thicknesses measured in mils. A mil is one thousandth of an inch. Recording tape is made in 1.5 mils, 1 mil, and .5 mils. High quality audio tape is usually 1.5 mils while cassette tape is frequently .5 mils. The thickness of the tape determines the strength of the tape, and most high quality studio recorders will not play .5 mil tape because the tension on the reels snaps the tape. The thickness of the tape is also a consideration in

Magnetic material (this side toward tape head)

Base of mylar or polyester

Exhibit 3.15.   Edge View of Magnetic Tape

Exhibit 3.16.   Tape Recorders used in Radio Broadcasting
(*above*) Ampex AG–440C four channel studio open reel recorder. Photo courtesy of Ampex Corporation. (*opposite, top*) Standard single-play audio tape cartridge. Over 99 percent of all U.S. radio stations now use tape cartridges due to ease of storage and automatic recue features. Photo courtesy of International Tapetronics Corporation (ITC). (*opposite, bottom*) Technics RS–M85 studio cassette recorder manufactured by Panasonic. Photo courtesy of Panasonic.

recording fidelity, since a thin tape is more subject to the recorded signal penetrating more than one thickness of the tape and causing what is called *print-through*—the transfer of the magnetic field from layer to layer of the tape on the reel.

Open Reels, Cassettes, and Cartridges

Tape recorders are available in three standard formats—open reels (reel-to-reel), cassettes, and cartridges. (See Exhibit 3.16.) Each type of recorder has different production functions.

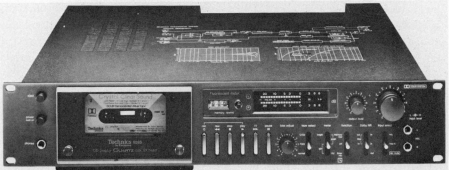

The open-reel recorders are considered the best for radio production because they can be easily edited. This means that the tape can be removed from the reels at any point and tape can be cut or added. (Complete editing instructions are provided in Chapter Six.) Most moderately priced open-reel recorders provide excellent fidelity. The large studio open reels are always used in full-production recordings.

The cassette tapes are composed of two tiny reels — a reel loaded with tape and a take-up reel, both enclosed in a plastic shell. The cassette tapes are much more portable than are the open reels, since both reels of the cassette are contained in one plastic case measuring approximately two-and-one-half by four inches. Cassette recorders are excellent for field interviews, since the light weight of the cassette recorders and the convenience of the small cassettes make these recorders the most portable.

Generally speaking, cassettes provide less fidelity than do most open reels because the tapes move at a slower speed than does tape on most open reel recorders and, partially, because the tape measures one eighth of an inch in width compared to one fourth of an inch in width for open-reel recorders. The cassette, in most cases, can be turned over and both halves of the tape used, which means that the recorded signal is being placed on a magnetic material that is slightly less than one sixteenth of an inch. In general, the smaller the recording space, the less the fidelity of the recording. Engineers have worked to solve the problems with fidelity on cassette recorders and have developed cassette systems that can provide excellent fidelity. Because of these technological advances, cassettes are becoming more popular, but like other products, there are qualitative differences in cassette recorders. If inexpensive, they may lack the ability to produce good fidelity and would not provide broadcast quality sound.

Cassette recorders are not considered to be production recorders because cassettes are difficult to edit. Frequently, when the cassettes are used for field recordings, the tapes are dubbed to "carts" or open reels and sometimes put through filters and equalizers to improve the sound fidelity.

Cartridge tapes (carts) are different from open reels and cassettes, since the cartridge is composed of one continuous loop of tape. The same length of tape circles around one reel. The cartridges, like the cassettes, are encased in plastic that measures approximately three by five inches. Cartridges are not generally considered to be production tools because they are not easy to edit and they are loaded with just enough tape to provide twenty seconds to five minutes of recording time. Cartridges are generally used for inserts into a program — opening or closing music, advertisements, public service announcements, short interviews, and sound effects.

Cartridges have greatly improved the job of the audio board operator because they are automatically cued. A tone, which is not audible to the human ear, is placed on the cartridge when the recording is made. This tone serves as a signal for the recorder in recueing the tape. If a piece of music is to be used on the open and close of a program, the music can be recorded on a cart, played for the program open, and the cart machine will automatically cue the cart back to the beginning of the music so that the same music can be used for the program close.

Carts and cart machines are an important part of the automated systems found in many stations today. The cartridge carousels are loaded with the carts containing the advertisements and the public service announcements (PSAs) to be aired for a particular program segment. The computer controls of the automated systems are programmed to play the various carts at precise times. With this new computer-controlled equipment, the station can be sure that ads are played precisely when the purchaser expects to hear them, and employees are not required to cue and play each ad manually. Cart machines have revolutionized station operations, making the job of the audio operator easier and less tense.

Each recording machine records and plays at certain speeds. That is, the tape is transported across the heads of the machine at 1⅞ inches per second (ips), 3¾ ips, 7½ ips, 15 ips, or 30 ips. The more slowly the tape moves across the record head, the longer the tape can be used for recording. Obviously, a tape that is recording information at 1⅞ ips is going to require less tape for the same recording than is a tape moving across the record heads at 30 ips, or 15 ips, or 7½ ips. (See Exhibit 3.17 for information on tape length and record/playback time.)

Most studio reel-to-reel machines record at 7½ and 15 ips, while most cassette recorders record at 1⅞ and some professional model cassettes record at 1⅞ and 3¾ ips. As a rule, recordings using the most tape per second will provide the highest quality sound reproduction, since there are more magnetic domains being magnetized to reproduce a signal.

Exhibit 3.17.
Recording Time for Various Tape Lengths at
Varying Recording Speeds.

Tape Length in Feet	Tape Speed in inches per second (ips)			
	1⅞ ips	3¾ ips	7½ ips	15 ips
150	15 min.	7½ min.	3¾ min.	1⅞ min.
300	30 min.	15 min.	7½ min.	3¾ min.
600	1 hour	30 min.	15 min.	7½ min.
900	1½ hours	45 min.	22½ min.	11¼ min.
1200	2 hours	1 hour	30 min.	15 min.
1800	3 hours	1½ hours	45 min.	22½ min.
2400	4 hours	2 hours	1 hour	30 min.
3600	6 hours	3 hours	1½ hours	45 min.
4800	8 hours	4 hours	2 hours	1 hour

Full Track and Multitrack Recording

Another distinguishing feature of tape recorders is their ability to record various channels of sound. The type of electromagnet used for the recording head determines the recording format that can be made. These head arrangements include full and half-track monaural (single-channel sound), two-channel stereo, four-track stereo, and four-channel or quadraphonic formats.

All early recorders were full track. This means that the electromagnetic head covers the entire width of the quarter-inch tape with a recorded signal. This is shown in Exhibit 3.18.

A half-track or two-track monaural recording places the signal on one-half of the tape, allowing the tape to be turned over for recording on the other half of the tape. Each reel of tape can contain twice as much information as can a reel of tape recorded on a full-track machine. (See Exhibit 3.19.)

A two-track stereo uses the entire width of the tape to record two signals — one to be played through the left stereo channel and one to be played through the right stereo channel. Exhibit 3.20 demonstrates two-track stereo.

Four-track stereo uses half of the tape to record a left and a right channel and the tape can be flipped over to record another stereo recording on the "other side." (See Exhibit 3.21.) Four-track stereo was a major technological breakthrough that opened the door for advances in the recording industry because it allowed more information to be placed on smaller and smaller portions of the tape.

Cassette recorders use a variation on the four-track format shown above. Instead of using tracks one and three or two and four for a stereo program, they use tracks one and two or three and four[2] as shown in Exhibit 3.22.

Eight-track stereo is the recording system that we have become familiar with in our automobile cartridge systems. The eight-track cartridges allow four musical programs to be recorded on one cartridge and offer a high level of fidelity while effectively utilizing the tape space to provide the greatest playback time. For program 1, the heads read tracks one and five; for program two, tracks two and six; for program three, tracks three and seven; for program four, tracks four and eight. Eight-track stereo is demonstrated in Exhibit 3.23.

The recording industry has become very sophisticated in maximizing the space on a magnetic audio tape. Some recordings are now made with sixteen, twenty-four, thirty-two, and even sixty-four tracks, allowing maximum control in blending a variety of sounds — especially music. Many tapes are made by mixing the musical elements — vocal, then rhythm, then piano, then background music, and so forth. The objective is to produce a recording that blends the various audio elements into the exact unity of sound that is desired.

These recording systems are much more sophisticated than the equipment

2.   This allows stereo recordings to be used on monophonic cassette recorders where the two stereo channels get mixed into one in a half-track monaural format.

Full Track Monaural

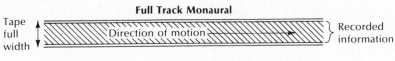

Exhibit 3.18. Full-Track Monaural Recording

Half-track Monaural

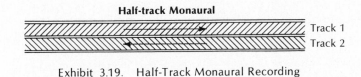

Exhibit 3.19. Half-Track Monaural Recording

Two-track Stereo

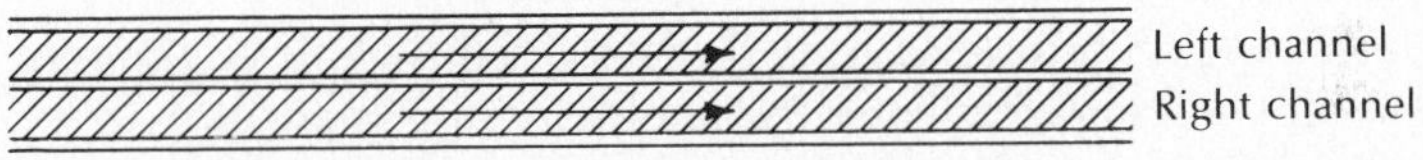

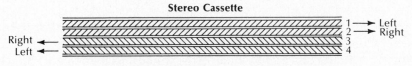

Exhibit 3.20. Two-Track Stereo Recording

Four-track Stereo

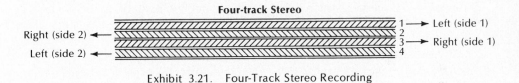

Exhibit 3.21. Four-Track Stereo Recording

Stereo Cassette

Exhibit 3.22. Stereo Cassette Recording

Eight-track Stereo Recording

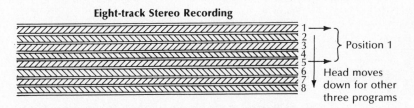

Exhibit 3.23. Eight-Track Stereo Recording

that is likely to be found in a standard radio studio. A student interested in multitrack recording should take courses that specialize in sound recording or should serve an internship at a recording studio.

Selecting a Recording System

A recording system can be as simple or as complex as the demands of the particular recording and the resources available to accomplish the task. The important questions to be asked when selecting a recording system are:

1. What is the exact nature of the recording project?
2. Where will the recording be made?
3. How will the recording be used?
4. What is the length of the recording?

Answering these questions will dictate the type of equipment required for the project. A recording of a two-hour lecture may dictate a good quality cassette recorder, especially if the lecture is across town and transportation is by foot. In this case, portability of equipment and recording supplies is one of the most vital considerations.

If a recording is to be made of the New York Philharmonic in the university performing arts center, much more sophisticated equipment is demanded. Stereo recorders with high quality microphones should be used. Equipment will most likely be transported in a van.

If a radio drama is recorded in the radio studio, a full-track monaural recorder may be the best choice since open reel, full-track provides editing flexibility. Ease of editing is a crucial consideration when producing a radio drama.

If thirty seconds of opening theme music are to be recorded for a radio program and the same thirty seconds of music are to be used on the program close, a cart machine will serve this purpose. The music should be recorded on a forty-second cartridge, which the cart machine will automatically cue for the close of the program.

Whether a production is to be made in the studio, in a remote van, or in the field, no decision should be made about the equipment needs before answering the above four questions. By understanding the features of a particular recording system, equipment can be selected that will be optimum for the particular project.

TURNTABLES

The Mechanism

The turntable is, unlike the recorder, a playback only device. The grooves of the record contain a mechanical equivalent of a sound. The stylus is led down a "wiggly" road where the bigger the deviation from a smooth road the larger is

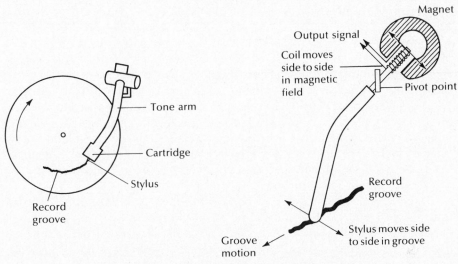

Exhibit 3.24. The Turntable and Cartridge System

the sound signal, and where the number of "wiggles" per second determines the frequency of the sound signal. The conversion of this mechanical motion into electrical signals relies on our old friend magnetic induction.

The moving stylus is connected to a coil that is suspended in a magnetic field. As the coil "wiggles" in the field, a current is induced in the coil, and this current has the same amplitude and frequency characteristics of the grooves in a record.[3] In the case of a stereo recording, each side of a groove contains "wiggles" that represent one of the two channels of sound. A special stylus and cartridge (the stylus housing the coils and the magnet) are necessary to decode the signals from the two sides of the groove separately.

Exhibit 3.24 provides a pictorial characterization of the stylus pickup. The exhibit shows a fixed magnet and a moving coil, but some cartridges use a fixed coil and a moving magnet. Any motion that results in a changing magnetic field will produce a changing current in the coil — it doesn't matter if it is the coil that moves or the magnet.

In order to track the small and rapid oscillations of the record groove, the stylus and its suspension must be very compliant — they must respond easily to very small forces. For this reason, the operator must be careful when handling the tone arm so that the stylus and cartridge are not harmed.

As in the case of the tape recorder, speed accuracy is directly related to pitch accuracy. A 1 percent variation in speed will be heard as a "warbling" of a sustained tone. Further, since records are recorded at very precise speeds of $33\frac{1}{3}$ or 45 revolutions per minute (RPM), the turntable should spin the record at just these speeds.

---

3. The grooves in the record were originally produced by the reverse of this process. The cutting stylus is driven by a varying magnetic field caused by the audio signal in a coil. The cutting stylus thus carves out a mechanical replica of the sound signal in the master disc.

Turntable Operations

The turntables designed for professional radio operations function much as manual single-play record players found in most homes. The turntables are designed to be durable, but special care should be given to the pickup arm and the stylus. These should always be treated gently and should be placed lightly on records and removed carefully. Nothing should ever be placed on the turntable platter except records; the platter has been balanced for proper rotation and heavy objects placed on it may damage the balance. The platter should always remain clean, since the records will collect any dust on the platter. Most turntable platters have some form of dust cover, and this should always be placed on the turntable when not in use.

   The worst enemy of the record and the turntable is ordinary dust. The particles of dust that float around in the air and settle on everything are small enough to fit in the groove of a record. Even worse, the sliding of the stylus in the groove produces static electricity on the record, which attracts dust. The problem can be minimized by keeping the platter and turntable top clean at all times (under a dust cover when not in use). Records should always be placed in their jackets after use. A variety of record cleaning materials and antistatic devices are available to help keep records clean. The problems caused by dust in the grooves include rapidly worn record grooves and increased stylus wear, but, worst of all for radio production, a noisy sound signal. Every dust particle can become a pop, and a scratch can be heard as a regular pop every time it comes around. For a particularly dirty and noisy disc, washing with a mild soap and a spray rinse may be necessary.

   The turntables found in the radio control room are easy to operate and convenient to use. Many have automatic cue devices and remote controls to further the convenience of operations. The beginning broadcaster should experiment with using the turntables. But at all times, careful treatment should be a standard part of the operator's procedure. (Digital turntables are not yet used in radio broadcasting. A discussion of digital audio appears in Chapter Ten.)

SPEAKERS

The intended goal of the radio signal is for someone to hear it. A speaker or speaker-like device is needed (headphones, for example) to convert the audio signal to a real sound wave. Nearly all speakers work just the opposite of a dynamic microphone. Again, magnetic induction is the underlying principle of operation. Since the speaker must move a large quantity of air to produce loud sounds, there is need for a large magnet and a coil with lots of windings so that the magnetically produced forces are large enough to move a big diaphragm.

   The output of the amplifier is connected to the speaker coil; the varying current in the coil causes the coil and the cone to which it is attached to move in

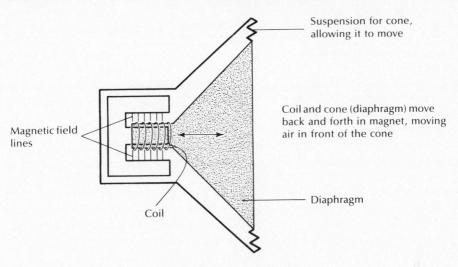

Suspension for cone, allowing it to move

Coil and cone (diaphragm) move back and forth in magnet, moving air in front of the cone

Magnetic field lines

Diaphragm

Coil

Exhibit 3.25.   The Speaker

and out of the magnet in the speaker base. This motion is large for large currents and thus moves a lot of air in front of the cone (this causes loud sounds). Also, the cone moves in and out with a frequency that is the same as the electrical signal; thus, the number of sound waves per second (the pitch of the sound) is the same as the frequency of the audio signal. Exhibit 3.25 demonstrates the speaker and how it works.

Frequency response is an important consideration for the speakers. Large speakers move lots of air and are good for producing long wavelengths (low frequencies). But a large cone doesn't move very quickly. For this reason, a small cone is used for reproducing the high frequency sounds. In most good speaker systems, this is carried one step further and at least three speakers are used: 1) a *woofer* to reproduce the low frequencies (20–500 Hz), 2) a *midrange speaker* to reproduce the middle frequencies (500–3,000 Hz) and 3) a *tweeter* to reproduce the high frequencies (3,000–20,000 Hz).

The size of the speaker goes down proportionately: a woofer ranges from 8–15 inches in diameter; a midrange speaker ranges from 4–8 inches in diameter; a tweeter is usually 2½–3½ inches in diameter. A special circuit is built into the speaker called a *crossover network* that routes the appropriate parts of the signal to each speaker. Such a speaker system will typically have a frequency response of 20–20,000 Hz, ± 3 dB.[4] In the case of a stereo system (either at the receiver or on the studio monitors), one has two such speaker systems — one for each channel.

4. A ± 3 dB means that in the range of 20–20,000 Hz, the curve never deviates from flatness by more than 3 dB up or 3 dB down.

MIXERS: MODEST OR "SUPERSTOCK"

Function and Operation

A mixer, which is also called a console or a control board, is a piece of equipment through which sound sources can be heard singly or several sound signals can be blended together or mixed. The purpose of a mixer is to allow two or more signals to be combined with separate control of the levels of the signals.

In general there are two kinds of mixers—broadcast mixers and full-production studio mixers. The broadcast mixer is usually less expensive and has fewer source inputs than a full-production board. Modern audio production studios often mix as many as twenty-eight to thirty sound sources, whereas in broadcasting, only a maximum of four channels (as in quadraphonic broadcasts) can be aired at once. A radio station that produces many of its own programs may have a production-type console, but many radio stations will broadcast from a very simple console that has only a few source inputs.

The Mix-Down

A process known as *mixing down* is used to blend the various sound sources into the desired signal. This process involves controlling the relative levels of all the input channels in some combination that produces either four final channels of sound (quadraphonic) or two final channels (stereophonic) or one mixed signal (monophonic).

The mix-down process can be very artistically complicated if one begins with many tracks of recorded signals. A multichannel mixer is technically no more complicated than a two-channel mixer. The number of signal combinations and the level at which each should be recorded makes the task of producing a mixed-down signal not a technical problem but rather an artistic problem.

If the audio operator is mixing music, for example, the piano may be one sound source, the horns may be another, the drum another, the lead singer another, the background singers another, and so forth. The audio console operator must decide how much of each sound is needed in the final signal. The mixer allows an infinite number of possible blends, one or more of which will sound "right" to the producer.

Each console has potentiometers (pots) through which volume levels for the various sound sources are controlled. Exhibit 3.26 shows an audio console. The sound inputs that can be fed into an audio mixer and amplified are microphones, turntables, tape recorders, cart machines, or any piece of audio equipment that is wired or patched into the console.

The Volume Unit Meter

The volume unit meter or VU meter is located in the center of the audio console and is labeled with two scales—decibel and percent modulation. The decibel

Exhibit 3.26.   The Audio Console
The Harris M–90 audio console is completely modular and offers a wide choice of
operating characteristics. Photo courtesy of Harris Corporation.

scale is labeled from −20 to 0 and then from +1 to +3. The percent modulation scale is labeled from 0 to 100. Exhibit 3.27 shows a close-up of a VU meter. Beyond 0 dB or 100 percent modulation, the meter is painted red. A signal that keeps the needle in the red is overmodulated and is generally a distorted signal.

The easiest check of proper sound levels for the beginning broadcaster is to monitor the percent modulation scale. A good reading on this scale is when the VU needle oscillates between 60 and 100 percent modulation. A reading of between 0 and 40 percent would be low and is said to be "in the mud" because the sound is not clear. A reading of over 100 percent modulation is "in the red" and is said to be "overdriving," causing distortion in the sound.

With any sound source, the volume will vary greatly from moment to moment. In human speech, for example, when an individual pauses briefly, the VU

Exhibit 3.27.   The Volume Unit (VU) Meter

The volume unit or VU meter is marked in percent modulation (0–100) and in decibels (−20–+3). A good VU reading rides in the 60–100 percent modulation range. The board operator should be careful that the needle never stays in the area over 0 dB or 100 percent modulation.

FADE IN: Bring the sound up slowly from a zero reading to a good volume level.

FADE OUT: Take the sound out slowly from a good volume level to a zero level.

FADE UNDER: Take the sound down from a good volume level to a low level. The fade under is generally used when one sound source is added to another. Voice is frequently added to music, for example. When an audio announce runs over music, the music should be in the background; the voice and music should not be in competition with each other.

IN CHANNEL: The pot control switch is in the air or broadcast channel.

OUT OF CHANNEL: The pot control switch is not in the air channel. This means that the channel selector is in the neutral or cue position.

POT DOWN: Take the volume level down for a particular pot. This is a frequent command when the level is high enough to cause distortion.

POT UP: Bring the volume level up for a particular pot. This is a frequent command when the level is so low that the sounds are unclear and muddy.

RIDE GAIN: To control the VU level for a sound source that is constantly changing. When riding gain, the board operator generally keeps a hand on the pot at all times.

SEGUE: When two sounds connect with no pause between them. A segue is used when one piece of music begins at the exact moment that another piece ends.

Exhibit 3.28.   Basic Definitions for the Audio Board Operator

level will drop to the 0 modulation level and when the person begins speaking again, the level may be momentarily over 100 percent modulation. The board operator should constantly monitor the VU level because levels may vary greatly from one sound source to another. The board operator should be especially careful when combining two sound sources, since the sounds from two sources can easily send the needle into the red.

Using the Control Board

Operating an audio console can be a frustrating experience for a beginning broadcast student because of the many knobs and switches that must be properly set for board operations. Every board functions similarly, and once the broadcaster has mastered one board, any other board will require only an introduction. The only way to learn proper board technique is to practice in the control room. Once the board operator learns the basic switches and knobs for controlling sound sources, the operator can begin to mix sounds. Exhibit 3.28 provides some definitions that will be useful to the audio board operator.

PATCHING

In many stations, all but the most unusual of sound sources are prewired into one of the available positions on the audio console; other stations require sound sources to be patched into the board. Patching is the process of connecting one piece of equipment to another. In the radio control room most patching involves connecting a sound source such as a turntable, a cart machine, a microphone, or a tape recorder into an audio console.

Connecting one piece of equipment into another is done by using a patching cord and a patching panel. The patching cord is a short cable with an identical plug or plugs at each end. The patching panel itself is a display of numerous holes (each labeled) in a metal rack. Exhibit 3.29 shows a typical patching panel.

In general, patching usually requires using the cable or patching cord to jack into the sets of holes that are labeled for the two pieces of equipment you

Exhibit 3.29.   The Patching Panel
A patching panel showing several connections.

are connecting—connecting a turntable into pot #3 on the audio board, for example. Sometimes the patching process is complicated by the requirement that the two pieces of equipment you wish to connect must be electrically matched to each other, i.e., the impedance must be matched.

Impedance (as discussed with microphones) is the resistance to the flow of alternating current in an electrical circuit. Radio equipment is characterized as high impedance or low impedance. A mismatch of a piece of high impedance equipment to a piece of low impedance equipment may result in a power loss or in a loss of sound fidelity. In most stations, the engineer has prepared a list of the various pieces of equipment that must be patched, and the engineer should have indicated the steps necessary to match properly the various pieces of equipment through the patching panel.

In state-of-the-art radio stations, the manual patch panel described above is beginning to be replaced by all-electronic switches, in which the patching is done by selecting a button at the audio console. In this case, the patching controls become part of the control board.

This discussion of patching is quite general and serves only to point out the basic ideas of patching. Every radio control room has its unique situation and will need to be studied on an individual basis.

SOUND-SHAPING DEVICES

Improving Sound Quality

As discussed earlier in this chapter, the shape of a sound can be measured and demonstrated as a frequency response curve. Through sound-shaping devices, the characteristics of a sound can be altered to achieve the particular sound that an audio operator wants.

If a sound is deficient in a particular part of the frequency spectrum, that part of the spectrum can be boosted to create the desired sound. If there is a low frequency hum in the sound, some of the low frequencies can be eliminated to reduce the hum. These are some of the ways in which an audio operator can change the shape of a sound. The devices used to accomplish these feats are referred to as frequency filters and equalizers.

Filters

Audio filters are devices used primarily for eliminating various frequencies in a sound. Several types of filters are found in the radio station and may be useful in correcting errors or problems with a sound signal. A filter may be of the active or passive type. An active filter is one with compensating amplification, and a passive filter is one without compensating amplification. Whether the filter is active or passive, it will be one of three basic kinds: 1) low-pass, 2) high-pass, or 3) notch.

The low-pass filter has an upper frequency cutoff (usually fairly sharp) and will pass all frequencies below that cutoff. The high-pass filter has a low frequency cutoff and will pass all frequencies above that cutoff. The notch filter allows the operator to reject a "slice" of the audio spectrum centered at some selected frequency. The bandwidth determines the size (in Hz) of the "slice" being rejected.

Some examples of the use of these filters are:

1. Low-pass filters are useful in eliminating hiss or white-noise sounds, particularly in recorded signals.
2. High-pass filters are useful in eliminating hum or rumble-type noises.
3. Notch filters are particularly useful if a well-defined noise appears in a recorded signal, such as the whine of a motor.

Exhibit 3.30 shows the effect of each of these filters on a relatively flat output spectrum.

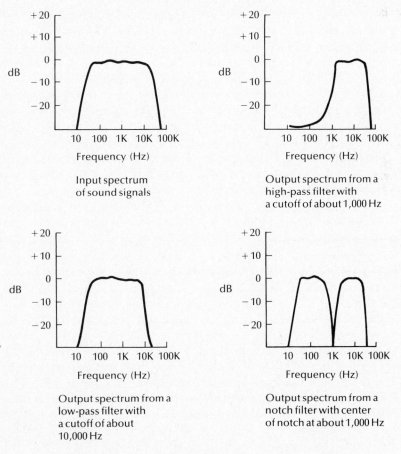

Exhibit 3.30. Effects of Three Kinds of Frequency Filters

Careful use of filters, alone or in combination, can cure numerous ills in sound signals. An audio operator may also want to do more to a given signal than just reduce a portion of the sound spectrum. In this case, a more general sound-shaping device called an *equalizer* is appropriate.

Equalizers

An equalizer gives the audio operator the ability to view the audio spectrum as a series of frequency slices—usually based on octaves—over which the operator can have independent control to enhance or attenuate the sound level (by as much as 15 dB). The equalizer can be a very powerful tool for correcting a recording that is, for example, generally lackluster on the high frequency end, or maybe too brassy. A microphone with a poor frequency curve could be at least partially equalized to a more normal frequency response with an equalizer. Exhibit 3.31 shows a typical audio equalizer.

Equalizers allow simultaneous enhancement and attenuation of any of a number of sections of the audio frequency spectrum. Exhibit 3.32 shows pictorially the sort of effects one can get using an equalizer which divides the spectrum into five frequency segments.

A poor recording can never be made into an excellent recording by filtering and equalizing, but these devices can correct for various problems with microphones, playback systems, and so forth. Their use, however, demands a good deal of familiarity with the specific characteristics of filters and equalizers available in a given studio. The best way to become creative in the use of these tools is to use them often enough to be aware of their properties. Out of such experimentation will come an appreciation of the power of these devices to assist in the production of excellent radio programs.

Noise-Reduction Circuits

In addition to frequency filters and equalizers, which are frequency sensitive sound-shaping devices, there are also amplitude sensitive sound-shaping devices. These fall under the general description of noise-reduction devices.

The simplest noise-reduction circuits, such as pop or scratch filters, work by eliminating or attenuating sounds that have a very fast "rise time" and a large amplitude. Suppose, for example, that a stylus is following a smoothly varying groove in a record and hits a rather large piece of dust. This event results in a very sharp "spike" of voltage output from the phono cartridge. Such a spike stands out clearly above the ordinary signal. The noise-reduction circuit is sensitive to this sharp "spike" and can distinguish it from the other normal sounds. The noise-reduction circuit functions by attenuating the signal for a very short period of time—long enough for the "spike" to be gone.

Notice from the above description that there may be real sounds (imagine

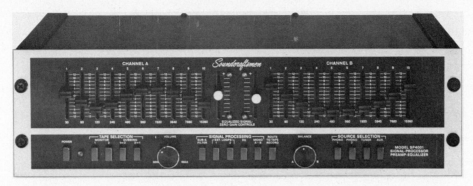

Exhibit 3.31. The Audio Equalizer
A typical studio-quality equalizer. Photo courtesy of Soundcraftsmen.

the crash of cymbals, for example) that resemble the pop sound of dust on a record. Any sound that begins abruptly (percussive instruments in general) and has a fast "rise time" will be somewhat affected by such devices. Thus, even though these noise-reduction circuits often do a great deal to improve a sound, they can have adverse effects and should be used with this knowledge in mind.

Noise-reduction circuits can be much more complex than those just described. The most sophisticated ones combine frequency and amplitude analysis of the signal and respond in a variety of ways to the instantaneous signal. The most widely known of these is the Dolby noise-reduction circuit, whose main purpose is to correct for tape "hiss," heard predominantly in low amplitude signals played back on a tape recorder. A detailed discussion of this and similar noise-reduction systems is beyond the scope of this book. For the interested student, a number of good advanced texts are listed in the bibliography at the end of this chapter.

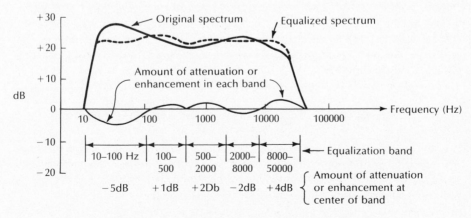

Exhibit 3.32. The Effect an Equalizer Can Have

RADIO EQUIPMENT
AND THE OPERATOR'S RESPONSIBILITY

This chapter has discussed the standard equipment in most radio stations. To use any piece of equipment properly, the operator must become familiar with the specific characteristics of the equipment. The first days in any radio control room should be devoted to mastering the control room setup and its unique equipment.

Proper production techniques should be learned for each piece of radio equipment. An understanding of the characteristics of the various pieces of equipment will enable the broadcaster to select the most functional and economical equipment to complete a broadcasting task properly and will dictate the best production procedures.

BIBLIOGRAPHY

Alten, Stanley R. *Audio in Media.* Belmont, Calif.: Wadsworth, 1981.

Everest, Frederich A. *Acoustic Technique for Home and Studio.* Blue Ridge Summit, Pa.: TAB Books, 1973.

Hasling, John. *Fundamentals of Radio Production.* New York: McGraw-Hill, 1980.

Johnson, Kenneth W., and Walker, Willard C. *The Science of Hi-Fidelity.* Dubuque, Iowa: Kendall/Hunt, 1977.

Mackenzie, George White. *Acoustics.* New York: Focal Press, 1964.

Nisbett, Alec. *The Technique of the Sound Studio.* 3d ed. New York: Hastings House, 1974.

Nisbett, Alec. *The Use of Microphones.* New York: Hastings House, 1974.

Oringel, Robert S. *Audio Control Handbook.* 4th ed. New York: Hastings House, 1972.

Rosenthal, Murray P. *How to Select, Use, Troubleshoot Cassette and Cartridge Recorders.* New Jersey: Hayden, 1972.

Seidle, Ronald J. *Air Time.* Boston: Holbrook, 1977.

Stevens, Stanley S., and Warshofsky, Fred. *Sound and Hearing.* New York: Time, 1965.

Traylor, Joseph G. *Physics of Stereo/Quad Sound.* Ames, Iowa: Iowa State University Press, 1977.

Whetmore, Edward J. *The Magic Medium: An Introduction to Radio in America.* Belmont, Calif.: Wadsworth, 1981.

QUESTIONS FOR REVIEW AND DISCUSSION

1. Sound waves can be characterized by wavelength and frequency. Within the range of human hearing, what is the wavelength of a sound at the low frequency end and what is the wavelength of a sound at the high frequency end?
2. Human hearing is generally within what frequency range?
3. If the pressure fluctuation of a sound wave is large, what characteristic will the sound have? If the pressure wave is small, the sound will have what characteristic?

4. On the decibel (intensity) scale, what is the range of human hearing without pain?

5. Sound waves reflected from walls or ceilings or furniture can have what effect on a recording?

6. What are some typical sounds at the following dB levels? 80 dB, 40 dB, 10 dB, 110 dB.

7. What is transduction?

8. What are the four basic methods of operation for microphones? Specifically, how do the ribbon and crystal microphones operate?

9. What are the advantages and disadvantages of the following microphones?
    Carbon
    Condenser
    Crystal
    Dynamic

10. What types of response or "pickup" pattern may a microphone have?

11. What is meant by frequency response in a microphone? What is meant by sensitivity?

12. Why may a music quality microphone not be appropriate for all recording purposes?

13. In selecting a microphone, what questions must be asked and answered to get the best equipment for a particular recording task?

14. Why is microphone balance both an art and a science?

15. What kinds of considerations should be made in selecting the best microphone placement for a particular recording situation?

16. What is the specific purpose of an amplifier in a piece of broadcast equipment?

17. Amplifier failure is usually indicated by what kinds of equipment malfunction?

18. In the record mode, how does the tape recorder work? In the playback mode, how does the recorder work?

19. How does the tape eraser or degausser work?

20. Evaluate the advantages and disadvantages of cart, cassette, and reel-to-reel recorders. What are the best uses for each type of recording equipment?

21. What is the thickness of standard broadcast recording tape? What standard tape lengths are used in professional broadcasting? What standard tape speeds are used for most radio station recordings?

22. What type of recording formats are available? Define: two-track stereo, full-track monaural, stereo cassette.

23. What important questions should be asked and answered in selecting a recording system?

24. How does the stylus function to decode the signal in the grooves of a record?

25. Why is dust said to be the worst enemy of the record and turntable? What can be done to protect the turntable system from dust damage?

26. How does the speaker function?

27. In relation to a broadcast speaker, define: woofer, tweeter, frequency range.

28. What are the frequency response limits of a good music quality speaker?

29. What is the purpose of a broadcast mixer? In percent modulation, what is a good VU reading for broadcast purposes?

30. In relation to a broadcast mixer, define:
     Fade In
     Fade Under
     Out of Channel
     Ride Gain
     Segue

31. What are the purposes of patching?

32. How do sound filters work?

33. In relation to sound-shaping equipment, define:
     Low-pass Filter
     High-pass Filter
     Notch Filter
     Passive Filter
     Active Filter

34. How do sound equalizers function? How can equalizers be used in broadcast production?

35. Why are noise-reduction systems used in broadcasting? Generally speaking, how do these systems function?

—————— Chapter Four ——————

BROADCAST COPYWRITING

Radio has been called theatre of the mind. The broadcast copywriter creates the stage, the set, the actors and the action.

The ability to write for the ear is necessary in acquiring a good job copywriting in radio. The types of writing required depend on the station and its format.

Many public broadcasting stations produce magazine news programs, interview programs, and feature programs and provide continuity for music programs. Many stations that broadcast popular music with an automated format produce only headline newscasts. A few stations produce commercials complete with sound effects and music, while other stations produce only short announce spots. There are many advertising agencies, also, that produce radio commercials and rely on broadcast copywriters.

The ability to write for radio can be a tremendous asset in job hunting. Today's station managers are looking for individuals who can operate the audio board, write copy, and effectively deliver the copy over the air. The more broadcast skills a student can demonstrate, the more likely that that person can get a good job in the broadcasting industry.

WRITING FOR RADIO

Writing for the Ear

What makes writing for radio different from any other form of writing? Primarily, radio copy must be written for the ear, since there are no visual stimuli to augment the spoken word. The words and the oral delivery must create images in the minds of the listeners. This is why radio is frequently called "the theatre of the mind."

Another complication for the radio copywriter is that the medium is used as a secondary activity by the listener. That is, the radio listener listens while driving an automobile, preparing dinner, working in an office, reading the newspaper, studying, or any of hundreds of other activities. Radio copy must catch the attention of the listener and must be easy to comprehend. Because radio listening is a secondary activity, there are certain rules that govern copy written for this medium.

Rules for Radio Copy

Sentences must be short. In radio copy, sometimes a phrase can be used instead of a complete sentence. "Stormy weather throughout the Midwest," "Complete details at ten," are examples of phrases that convey the idea of complete sentences. The radio listener gets the message. Complex sentences will not work for radio. Colons and semicolons are almost never used, and good copywriting should require few commmas.

Use the minimum words necessary to convey your ideas clearly. In the radio industry, time is very important. A commercial spot may take as few as ten seconds; a complete newscast may be allowed as little as two minutes; a news capsule may be given in fifteen seconds. With short blocks of time, there is no room for wasted words. Use as many words as necessary to convey your message correctly, but remember that the radio listener is probably doing something else while listening and is not concentrating enough for overly wordy copy. The best radio copy conveys the message in the minimum number of well-chosen words.

Make your copy conversational when possible and appropriate. A conversational quality is achieved in two ways: 1) by using contractions whenever possible and 2) by using the personal pronouns *you* and *your* whenever possible. Radio is a personal medium; it is a companion for the listener. The copy should help to create an environment of intimacy and should add a personal touch with which the listener can identify.

Note that this rule says *whenever possible and appropriate.* Most items in a newscast may sound inappropriate when presented in a personal manner, but some news items are appropriately presented with personal pronouns.

> "Your electric bill will be higher in June. The City Council voted last night to increase the electric rates."
>
> "Cover your tomato plants tonight. A hard freeze is forecast."

These are examples of the personal touch that can be introduced into informational copy. As a broadcast copywriter, you must decide when the personal touch enhances your message and creates a familiar and intimate environment for your listeners.

Use the active voice. The passive voice lessens the vitality of the copy. The active voice expresses immediacy while the passive voice expresses an action that has already been completed. Compare these two sets of sentences.

> "The police arrested a suspect at 10:00 AM today. The suspect remains in custody and is being questioned by the police." (*Active voice*)
>
> "A suspect has been arrested by the police. The suspect was placed in custody and the police have begun their questioning." (*Passive voice*)
>
> "A storm ripped through the city last night." (*Active voice*)
>
> "The city was hit by a storm last night." (*Passive voice*)

The active voice conveys immediacy; the active voice is stronger than the passive voice and adds vitality to your story.

Avoid complex words that the listener is unlikely to understand or to expect. Since radio listening is a secondary activity, the listener doesn't have time to ponder complex meanings. Broadcast copy should be familiar and the meanings of words immediately clear. *Expunge* is a good word, but your listeners are more familiar with *erase*. *Forgather* may be alright in prose writing, but your listeners are more familiar with the verbs *meet* or *assemble*. Overly complex language will cause your listeners to tune out and turn off.

When introducing individuals in your copy, the general rule is to give the person's title first. The title immediately identifies the individual for the listener. Since radio listening is a secondary activity, the title of the person may attract the listener's attention.

> "President of the teachers' association, Rose Smith, said . . ."
>
> "Chief White House aide, Bob Smith, reported . . ."

If the person's title is very short, the title may go after the name.

> "Bob Smith, Police Chief, . . ."

The better style, however, is to identify the person before giving the person's name. The title helps the listener to identify the news source immediately.

In addition to the above rules for radio copy style, there are also rules for preparing the copy. The copy format must be easy to read and must contain the essential information for the air personality. The simple rules listed below provide the best format for radio copy.

Radio Copy Format

Radio copy should always be typed and should be double spaced. Handwritten copy is unprofessional and may be impossible to read under the pressure of a live broadcast.

Ample margins should be provided at top, bottom, and sides of each page. One-and-one-quarter inches is about the proper space for margins on all sides. This gives the announcer room for notes and makes the copy easy to read.

Place essential information at the top of the copy in all capitals. At least one word to identify the copy, the date, the initials of the person writing the copy, and the amount of time necessary to deliver the copy should be placed in the top left corner. EARTHQUAKE/FRIDAY 6–15–84/LJB/45 SEC. identifies the copy, the date, and the copywriter and provides the delivery time. Advertising copy may be labeled in similar fashion: DELANEY FOODSTORE COPY/6–15 –84/LJB/30 SEC.

For news copy a kill date is usually placed at the top of the page. KILL/FRIDAY 6-15-84/7P.M. tells the newscaster that the story should not be used past the date and time indicated. "The Secretary of State is en route to Africa" would no longer be appropriate for a 10:00 P.M. newscast if the secretary arrived in Africa shortly after 7:00 P.M. Advertising copy announcing store specials may also need a kill date. The point to be made here is that *all* broadcast copy should be appropriately labeled at the top of the page. The topic, the date, the writer, and the delivery time are essential pieces of information for all copy.

Use abbreviations in the copy on words that are meant to be read as abbreviations: NATO, YWCA, AFL-CIO. Never abbreviate words that are not meant to be read as abbreviations. Avoid using abbreviations for names of countries, names of individuals, names of months, and so forth. September is not meant to be read as Sept., for example, and Street is not meant to be read as St. Abbreviations that are not meant to be read as abbreviations may get the announcer into trouble.

Exercise special care with numbers in broadcast copy. The general rules are: write out numbers under 10 and use the actual numbers between 10 and 999. Write out hundreds, thousands, and billions—16,000,400,000 is difficult to read; 16 billion, 400 thousand is easier to read. Numbers should be rounded off for ease of delivery and ease of understanding. Instead of writing "nine-thousand, nine-hundred and 50 tickets" were sold, the figure can be rounded off to "approximately 10 thousand." When using numbers in broadcast copy, the writer's goal is ease of delivery and ease of understanding.

These rules cover some general principles for all kinds of broadcast copywriting. There are also some specific rules governing various kinds of broadcast copy.

ADVERTISING VIA RADIO

An Effective Medium

Radio advertising is very popular, especially at the local level, and advertising rates are generally reasonable and competitive. One of the big advantages of radio is that the target audience for a product can be more easily identified via radio than it can be via most other media. Since listeners tune into a station because of a particular format, the advertiser knows that teenagers can be reached on a Top 40 station, senior citizens can be reached on an all talk, all news, or beautiful music station. A high school sports program is likely to attract a young audience and a heavily male audience. With radio, the advertiser can pick and choose just the right audience for the product. Radio can be very effective for the advertiser, but effective advertising demands well-written copy.

Prelude to Writing

When writing a commercial for radio, the first step is to *think about the product you want to sell.* You need to ask and answer many questions about the product. These answers are crucial in deciding upon the focus and scope of the commercial.

The first step, then, is *think before you write.* Ponder the following types of questions before writing any commercial copy.

What exactly are the special features of this product?

What facts do you know about the product?

Does this product have advantages over similar products? Advantages of packaging, color, size, convenience, economy?

Why would the buyer want to purchase this particular product?

Are there special copy angles that could be utilized with this product—time of year the product will be used, location at which the product will be used, unique shape or size of the product?

After this thinking process, you should be able to list unique features of the product and list some possible copy angles. You should complete this first step by writing some pages of information about your product—facts, unique features, advantages over similar products, and some objectives to accomplish in writing your advertising message.

Visualizing the Target Audience

The second step is to *select a target audience.* The writer cannot begin to write before understanding to whom the copy is directed. Based on the facts you have learned in step one above, decide what group of individuals represents the greatest sales potential for your product. Those individuals that you believe are most likely to buy your product constitute your target audience.

As a copywriter, you should try to know as much about this audience as possible. Try to visualize your audience. How old are its members? What does the target audience like to do? How do its members spend their time? What values do they maintain? What are the goals and self-images of members of the target audience? What is their basic lifestyle? What are the aspirations and goals of members of the target audience?

Tying the Information Together

The next step is to *tie the information you have learned in step one to the information you have learned in step two.* That is, how can the special features of the product be tied to the lifestyle and aspirations of the target audience? As you make these connections between the product and the audience, you should begin to develop some specific ideas for your copy. You should now have some clear objectives for your copy.

You can understand how the product and the target audience are linked by thinking about some products that are currently advertised in the broadcast media. What are the target audiences for various toothpastes, and what unique features of the product are stressed for the different target audiences? One toothpaste advertising campaign[1] may be directed at a young adult audience concerned about its social life, sex appeal, or acceptance from friends and colleagues. Another toothpaste advertising campaign may be directed at the American family concerned about dental health and checkups and cavities. The unique product features stressed in the ads are directly tied to benefits for members of the target audience.

One of the most important jobs of the commercial copywriter is to link properly the product benefits with the needs, goals, desires, and lifestyle of the target audience. When you have the links between your product and your target audience properly established, you are almost ready to begin writing. However, there are several more preludes that must be completed before the final copy is written.

Narrowing the Focus

Remember to *keep your ad simple.* Narrow the focus. Limit your sales pitch to one concept. Again, think of ads currently running in the media. Think of ads for vitamins. The appeal is clear—this product will preserve and improve your health. Think of advertising for automobiles. One advertisement may stress the

1. An advertising campaign is composed of a series of advertisements designed to sell one product. The advertisements can be carried in the same medium—radio, for example—or they can be carried in several media at once. A new toothpaste may be advertised in magazines, over the radio, on television, and on billboards. These various advertisements constitute the advertising campaign. Generally, a campaign will stress just one or two unique features about the product. A campaign across media is most effective if the target audience is diverse in its lifestyle and/or media use.

economy associated with a particular car. Another advertisement may stress the luxury associated with another car. Yet another ad may stress the reliability of a third car. The focus is always clear—economy, or luxury, or reliability. Each advertisement attempts to get *one* idea about the product across to the potential buyer.

Gaining Interest and Attention

One of the most important jobs of broadcast copy is to *gain attention* since the audience is usually doing something else while listening. The job of gaining attention is not an easy one. How can the broadcast copywriter do this? Music, sound effects, unique voices, and the use of special appeals are all available to the copywriter. Some of the special appeals that the copywriter can use include appeals to the need for safety, need for love, need for ego satisfaction, need for security, and many other human needs.

One appeal that could be used to gain attention is an appeal to the listener's *need for safety*. Advertisements for smoke detectors do this, as do public service announcements advocating the use of safety belts in automobiles.

Another appeal that can gain attention is an appeal to the listener's *need for love*. Some food commercials attempt to link good food with love from family and friends—a good cook shows love and receives love through cooking for friends and family.

The appeal to the listener's *sense of ego* is frequently effective. Friends and colleagues will be impressed when they see you in this automobile, or wearing this brand of clothing, or when you use this credit card.

The *need for security* is frequently an appeal that gains the attention of the audience. Some travelers check companies remind listeners that if their travelers checks are lost or stolen, they can be reissued at branches of the company around the world. The ads appeal to the listener's need for security. Insurance ads also generally appeal to listener's need for security.

Threat appeals can also be used, but they should be used with discretion. Denture creams often appeal to the listener's fear—without a good denture cream, your dentures are sure to slip and embarrass you. Occasionally, threat appeals are used in advertising for laundry products—dingy clothes reflect poorly on you and may cause the loss of respect from acquaintances. Advertisers have been known to use the same kind of threat appeals with deodorants—you may offend others if you do not use this product.

The list of appeals and examples of appeals in advertising copy could go on and on. There are humorous appeals, appeals to the listener's desire for fun, to the listener's desire for adventure, and appeals to hundreds of other desires, aspirations, and needs of the target audience. The important point is that the radio copywriter must gain attention and must make the listener desire the product. This is done by appealing to a particular need or goal of the listener.

Motivating Action

The final step is to *get the listener to take action*. The job of the advertising copy-writer is to convince the listener to purchase the product. This is accomplished by demonstrating that the need you have shown in your ads can be satisfied through the purchase of your product. The listener can give and receive love through purchasing this cake mix. The listener can impress colleagues by purchasing this car. The listener can protect the family by purchasing this insurance. The final step is to link the action of product purchase to need met, goal reached, or desire satisfied.

The procedure then in developing a commercial spot for radio is:

1. Evaluate your product and its unique features
2. Visualize your target audience and evaluate their lifestyle, needs, and goals
3. Tie together the information about the product and the target audience and develop an idea showing that the product benefits the listeners
4. Narrow the focus of the ad to make the appeal clear
5. Gain audience interest and attention by the effective use of appeals combined with unique voices, music, and sound effects
6. Show that the needs, goals, or desires of the members of the target audience can be met by purchasing the product.

Testing Your Ideas

Most advertising agencies test their advertisements on a subset of the target audience. While you may not be able to assemble such a subset for testing, you can get the response of a few friends. If your commercial is intended for a young adult audience, have some of your colleagues listen to your ideas and solicit their comments.

Do not ask for comments, however, unless you are prepared to listen. Listen with an open mind. If a similar comment is made by several friends, that's a comment you probably want to take seriously. If all of your friends like a certain angle of your commercial, you know that you have at least one very appealing aspect in the commercial. If, however, your friends don't like certain parts of the spot, you may want to consider changes. Do not be argumentative. Even if you disagree with your friends' comments, consider them. As a copy-writer, you must learn to evaluate your work critically; this means that you want to consider seriously all criticisms.

ADVERTISING FORMAT AND CONTENT

A 60-second spot will require approximately 150 words, depending upon your use of sound effects, music, and the delivery rate of your announcer. A

30-second spot will require approximately 75 words, if the spot is predominately words, with few sound effects or musical bridges. A 10-second spot will require approximately 30 words.

Don't rush your commercial. Allow your listener time to think. Remember that some repetition is needed in a commercial. In a 60-second spot, a general rule is to mention the client's name at least three times. If an address is given, this should be repeated at least twice. In a 30-second spot try to mention the product name at least twice.

Because repetition is important in radio, a mail-order product for which the listener must telephone or write would probably not be advertised in less than a 60-second format. The product name and address or telephone number need to be mentioned several times. Also, the listener needs time to write down the address or telephone number. For these reasons, the amount of time devoted to a commercial needs to suit the unique demands of the product. For every commercial, the listener needs time to think, time to identify the product, and time to consider the need and the appeal presented in the spot.

Review the rules for broadcast copy listed in the earlier section of this chapter. Remember to keep the copy personal, use contractions when possible, and address the ad directly to your listener.

Remember also to be sensitive to the social environment. Avoid humor at the expense of groups who may be offended by your use of stereotypes or slurs. Women and ethnic/racial minority groups are very much offended by stereotypes.

Be careful to avoid charged political or social issues in your ad copy. These issues generally have polarized the American public and may also polarize the potential customers for your product. Some advertising campaigns have attracted attention for precisely the wrong reasons—offensiveness to a large segment of the target audience. You want your listeners to identify positively with your commercial, not negatively.

THE COPYWRITER'S JOB: PRODUCTION TECHNIQUES

Production techniques are considered in Chapter 6, but the copywriter must also make some production decisions when writing the copy. The copy should indicate when music is to be used; it should also indicate when sound effects should be used. The copywriter will indicate the type of music and the particular type of sound effects to be used.

Radio stations or audio production companies should have a music library and a sound effects library to enable the copywriter's ideas to be put into reality. On the script, the copywriter should indicate the relative volume of the music or effects and indicate where voice is to be mixed with music and effects. Examples of such copy, written by broadcasting students, are shown in Exhibits 4:1, 4.2, and 4.3.

NEIGHBOR'S NOOK/6–10–84/HC/60 SEC.

KILL/7–10–84

(Music up—William Tell Overture)
(Music under)
Announcer:

> Are you tired of rushing from one store to another trying to find that special gift for that special someone in your life? Well stop!

(Overture ends—fades into softer, more relaxed music)
(Music fades under announcer)
Announcer:

> You can save yourself time and effort by shopping the Neighbor's Nook. You can take your time and relax in our pleasant surroundings, but what's more important—you'll find exactly what you're looking for at Neighbor's Nook. You'll find a variety of silk flowers, candles, glassware, and kitchen utensils. To make your present extra special, the Neighbor's Nook carries a wide selection of Hallmark cards. Gifts are our specialty at the Neighbor's Nook and we'll help you find the gift that's just right for that special someone in your life.

(Music up and out)

Exhibit 4.1.   Sample Student Script for Sixty-Second Commercial

---

Public Service Announcement

TORNADO WATCH/WARNING/5–30–84/HC/60 SEC.

KILL/9/30/84

Wind/weather sound effects
(Sound effects under announcer)
Announcer:

> Most of you here in central Iowa are familiar with severe weather such as tornadoes, which can destroy homes and property and can take lives.
> Tornadoes and severe weather are most prominent during spring and summer months and usually occur in the middle or late afternoon.

(Sound effects completely out)

(continued on page 93)

Exhibit 4.2.   Sample Student Script: Public Service Announcement

(continued from page 92)

Announcer:

In the event of severe weather, this station will provide you with complete information.

You should know the difference between a tornado <u>watch</u> and a tornado <u>warning</u>.

A tornado watch is when weather conditions are such that a tornado <u>could</u> form. In a tornado watch, you should stay tuned for more weather information.

A tornado warning means that a tornado has been sighted. You should take shelter <u>immediately.</u>

During severe weather conditions, this station will broadcast continuously. You can count on WXXX for complete weather news.

HAM STEAK BACON/8–30–84/60 SEC.

(Bacon frying under announcer)
Announcer:

If you think all bacon is the same . . . try <u>listening</u> to it sometime. You'll find most brands of bacon sound like this:

(Loud bacon noises up and over for 3 secs., then down and out)
Announcer:
(Softer bacon noise up and over 3 secs., then down and out)
Announcer:

But <u>Ham Steak Bacon</u> sounds like this:

You see . . . Ham Steak Bacon is the <u>quiet</u> bacon. (SHHHHHH . . .) It makes less noise when you fry it because Ham Steak Bacon is leaner . . . it's meatier . . . it has less fat.

(All bacon noise fades out)
Announcer:

That means it tastes a whole lot better than most bacons. All in all a pretty good reason for you to switch from your brand of bacon to Ham Steak Bacon.

Of course . . . if you buy your bacon for its acoustical qualities, you may find our bacon uninspiring. It doesn't have the melodious staccato rhythm of those other brands.

But, if you buy bacon for its flavor . . . (MMMMMMMMMMMMMMMMMMMMM) . . . you'll <u>love</u> Ham Steak Bacon.

Ham Steak Bacon . . . you can <u>hear</u> the difference.

Exhibit 4.3.   Sample Student Script for Sixty-Second Commercial

At some stations and production companies, you will be both copywriter and producer. At other operations, you may only write the copy and someone else may actually produce the copy. In any case, the copywriter needs to have a clear idea about production techniques so that the copy lends itself to audio production.

Many hours go into the writing and production of a 60-second radio commercial. The commercial radio station survives on the selling of advertising time, so good commercials are the key to keeping the station open. Because the very life of the station is dependent on commercial copy, advertising copywriting is one of the most important occupations at a commercial radio station.

SPECIAL PROGRAMS: FEATURES AND COMMENTARY

Copywriting for radio may involve the writing of feature programs. Few commercial stations carry special feature programs, but most public broadcasting stations do carry various kinds. The variety of feature programs is almost limitless.

Some stations feature a regular personality who comments on people and events in the news. Other stations offer programming that comments only on unusual events in the news. Other stations may have sports features or a local personality who comments on local events or on nostalgic topics.

Writing feature programs can be one of the most creative assignments offered by a radio station. This is the writer's chance to inform, to entertain, to demonstrate wit, knowledge, and wisdom.

The procedures for feature writing are the same as for most forms of broadcast copy. First comes the research on your topic; next comes the preliminary draft, written with the idea of preparing a script for the ear—sentences are short, words are clear and well chosen, images are clear, and think time is provided for the listener. Next, the second draft is written and tested on some members of the potential audience. The final touches are put on the copy and it is ready to be recorded.

A sample script of a feature program is provided for you to study in Exhibit 4.4. This script was written by veteran radio personality, Jack Shelley, with his own underlining for verbal emphasis. The Shelley script is from a feature comment program produced weekly on WHO radio, Des Moines, Iowa. This script demonstrates good broadcast copy—sentences are short, words are well chosen, images are clear, and the message is personalized for the individual members of the audience.

Exhibit 4.5 presents an introduction to a three-part historical series produced for public broadcasting stations and for use in schools. Note that this piece uses fewer personal pronouns than does most advertising copy or copy

HOME TOWN NEWS COPY/JACK SHELLEY/11-1-80

How do you do again, everyone. This is Jack Shelley with the Home Town News.

Let's start out with a pair of Iowa vignettes . . . little pictures of the Iowa scene during the fall.

We start with the front page of the Hedrick Journal. The paper had a story about three people from Wisconsin who passed through New Sharon not long ago—in a covered wagon.

That's right: Elmo and Helen Lun and their friend Bill McBride were traveling in two covered wagons, drawn by ponies and mules. And they plan to clip-clop some 3-thousand miles from Wisconsin to Texas, traveling some 20 miles a day.

McBride and Lun built their own covered wagons; they live in them on the road, and they're equipped with gas heaters to keep the passengers fairly comfortable—something the pioneers wish they could have brought along, I'll bet. They even carry what was de-described as a "portable pony barn." They can set it up at various stops and use it to offer shelter for the animals that pull the wagons.

The three friends from Wisconsin have taken short weekend trips in the covered wagons previously, but this is the first time they've embarked on anything this long—3 thousand miles!

Still, I've got to admit that what caught my eye first about this story in the Hedrick Journal, was the photograph of the sign on the back of one of those wagons. Here's what the sign says: "This vehicle is powered by oats. Don't step in the exhaust!"

Now, let's switch to the Postville Herald, up in the far northeast part of Iowa. And let me quote Maxine Degarmo, who says she was riding in a car through a small town up there, in September. "Suddenly," says Maxine, "I gave a startled sound and asked my companion, 'Did you see that Christmas tree?' "

Exhibit 4.4.   Sample Feature: Commentary Script
Jack Shelley script for Home Town News, broadcast over WHO radio, Des Moines, Iowa.

SECOND GENERATION SUFFRAGIST: CARRIE CHAPMAN CATT
3-PART SERIES/LJB/2-82

(Lively classical music up for several seconds and then under entire announce)

1848 began the women's suffrage movement in the United States. Five women in Seneca Falls, New York, called a public meeting to discuss the rights of women. "We hold these truths to be self-evident: that all men and women are created equal" was the battle cry.

The vote was the most basic right that women had to gain. The participants at Seneca Falls put forth a resolution calling for the enfranchisement of women. This resolution began the 72-year struggle for women's right to vote.

These first leaders did not live to see their dream realized. Two generations of suffragists were needed before women won the right to vote in 1920.

This is the story of one of the most significant women in that second generation of suffragists. Carrie Chapman Catt devoted her efforts to the cause of women's suffrage. She soon became the . . .

(At close of announce, music up for several seconds and fade out)

Exhibit 4.5.   Sample Script for Historical Series

for use at commercial broadcasting stations. The copy is informational, but sentences are short and the writer uses the active voice.

Rarely does the commercial radio writer have an opportunity to produce an historical series or a major production script, though the public broadcaster and the individual working for an independent production company may. Should the opportunity present itself, you, as a copywriter, would want to be able to accept the challenge, since writing a feature program or an historical piece will give you a chance to express your creativity and could result in a production of special pride for you.

Too few stations take advantage of feature programs. A feature program can be presented by a local citizen who has a sense of the topics that interest the local community or by a member of the station staff. A regular five-, ten-, or fifteen-minute feature program that airs daily or weekly adds scope and depth to the regular program offerings and will help to build a large and loyal audience. Unique feature programming can set your station apart from every other station in the market. But feature programming requires a copywriter who can find interesting topics and who presents the topics with wit and sensitivity.

THE ROUTE TO BROADCAST COPYWRITING

Any individual interested in broadcast copywriting should prepare carefully for the job. The preparation for a career in broadcast copywriting includes:

1. Take all of the university courses that will assist you in a writing career—courses in grammar, creative writing, continuity writing, and writing for radio, TV, and film.
2. Write often. Write advertisements, media continuity, feature programs, short stories, and other types of writing that fit into your daily assignments and your scope of interests.
3. Submit your writing to others for criticism. Evaluate the criticism and learn from it.
4. Serve an internship or find a part-time or volunteer job that will allow you to use some of your writing skills.

The above kinds of preparation should improve your writing skills and put you in a competitive position in the broadcast job market.

BIBLIOGRAPHY

Field, Stanley. *Professional Broadcast Writers Handbook.* Blue Ridge Summit, Pa.: TAB Books, 1974.

Hasling, John. *Fundamentals of Radio Broadcasting.* Chapter 11, "Writing for Radio." New York: McGraw-Hill, 1980.

Heighton, Elizabeth J., and Cunningham, Don R. *Advertising in the Broadcast Media.* Belmont, Calif.: Wadsworth, 1976.

Hilliard, Robert L. *Writing for Television and Radio.* 4th ed. Belmont, Calif.: Wadsworth, 1984.

Lee, Robert, and Misiorowski, Robert. *Script Models: A Handbook for the Media Writer.* New York: Hastings House, 1978.

Malickson, David, and Nason, John W. *Advertising: How to Write the Kind That Works.* New York: Scribner's, 1977.

Maloney, Martin, and Rubenstein, Paul M. *Writing for the Media.* Englewood Cliffs, N. J., 1980.

McLeish, Robert. *The Technique of Radio Production.* Chapter 5, "Writing." London: Focal Press, 1978.

Orlik, Peter B. *Broadcast Copywriting.* Boston: Allyn and Bacon, 1978.

Willis, Edgar, and D'Arienzo, Camille. *Writing Scripts for Television, Radio, and Film.* New York: Holt, 1981.

QUESTIONS FOR REVIEW AND DISCUSSION

1. What makes writing for radio different from any other kind of writing?
2. Radio listening is called a secondary activity. What does this mean for the broadcast copywriter?
3. List and discuss the rules for writing radio copy.
4. What information should be placed at the top of any broadcast copy?
5. What is meant by a "kill date"?
6. How should numbers be written in broadcast copy?
7. What is meant by "target audience" for the advertiser?
8. Why is a target audience easier to identify via radio than almost any other medium?
9. Why is radio a good advertising medium?
10. What kinds of questions should you ask about the product before you begin writing an advertisement for the product?
11. What questions do you, as a broadcast copywriter, need to ask about the target audience?
12. How do you use the information you have gathered in your product and target audience research?
13. List four advertising campaigns that are now carried over radio or in any other medium. What is the target audience for each campaign? What major concept is stressed in the advertisements?
14. List four advertisements that now run in the mass media. What appeals are used in each?
15. What are the steps in writing advertising copy?
16. What are the approximate number of words required for a 60-second spot when little music or sound effects are used?
17. Why is repetition important in writing commercials for radio? In a 60-second spot, about how many times should a product name be mentioned?

18. Why is it important to avoid stereotypes and charged social/political issues in writing radio advertising copy?

19. What kinds of production information does the radio copywriter need on the actual script?

20. What kinds of special programs can be produced for radio?

21. How can feature programming help a radio station?

--------------- Chapter Five ---------------

THE RADIO NEWS PROCESS

*The impact of radio has increased steadily since the early
1920s, when listeners crowded around crystal sets to hear the
latest election returns. Moreover, the growing national desire
to "be informed," and be informed "on the run," has increased the importance of the medium as a prime dissemina-
tor of news.*

John Bittner and Denise Bittner,
Radio Journalism, 1977

Most Americans consider radio news to be a vital part of everyday life. Many
wake up to radio news by way of a clock radio. In fact, radio's largest audience
is from 6:00 A.M. to 7:00 A.M. when Americans are waking up and dressing for
work. Americans tune in the radio in the morning to find out what has happened
in the world overnight, to find out what the weather will be like for the day, and
to find out what will be happening in the local community and in the nation.

JOURNALISM VIA RADIO

The Immediate Medium

Radio has a distinct news advantage over every other medium — immediacy.
When a story breaks, Americans expect to get the details first on the radio. A
radio station may have news on the hour, on the half hour, or news five or six
times a day. Other stations may be all-news stations. A television station
generally has news much less often than a radio station. A newspaper has a
delay from the time of occurrence of a news event to the printing of the article
in the paper. Radio is the first medium to break a story, and it is generally the
medium that Americans count on for immediate details.

Radio is especially important for local news. Many small communities do not have a local TV station and may have only a weekly newspaper. Radio becomes the medium that informs the community on a daily basis.

Even in larger cities, people count on the local radio stations to inform them about events at the city council meeting, the zoning commission, the planning commission, and events at other civic and community meetings. Radio has always been a local medium and its news service to the local community is vital to the very life of the community.

Radio news also serves the local community in times of disaster. In an electrical storm or during a tornado warning, the transistor radio is always a companion as people take shelter. In times of disaster, most people count on the radio for a minute-by-minute report.

The Standard Newscast

Many news options are available to the radio station—regular newscasts, interviews, documentaries, minidocumentaries, feature stories, and feature news programs. However, the most common type of news program at the radio station is the regular newscast.

The news copywriter counts on a wire service for these regular newscasts. Most stations subscribe to an international/national news wire service. Associated Press (AP) and/or United Press International (UPI) are most likely to be found at local commercial stations. The printout should be gathered from the wire at least every hour and should be categorized according to international, national, regional, feature, sports, and weather news. You may want to start separate files for fast-breaking major news stories. The latest news goes in the front of each file.

Both AP and UPI offer a broadcast wire service that prints the news in broadcast copy format. The stories still need to be edited, however, for the particular news format at your station.

You may want to use only one paragraph from the lead story and several sentences from other stories, then prepare a brief sports summary from several sports articles. The best procedure is to type your newscast using all the proper broadcast copywriting techniques. This procedure will give you the greatest familiarity with your stories and will be insurance toward a good newscast.

If you work at a very small radio station where you wear many hats and serve as board operator, receptionist, announcer, and newscaster, you may have to use the "rip 'n' read" technique. This is a poor form of journalism where you merely rip the news off the wire and read it over the air. The rip 'n' read technique usually results in faulty delivery and gives the listeners the impression that the station doesn't care about news.

If you must rip 'n' read, try to follow a news category system. As you have time, rip the news from the wire and put it in a file or on a clipboard that is properly labeled—international, national, sports, weather, top story, and so forth. If you do not have time to type and edit your newscast, at least underline the portions of each article that you will read and get a clear organizational pattern in mind. A common pattern is to begin with the most important national/international story, cover other national/international news, cover regional and local news, sports, and weather. This could be varied with all national/international news at one newscast and all regional/local news at another newscast. The organization of the news is dependent on the station's news policy.

The Station News Policy

The scope and coverage of news at any given station depends on the station's news policy. A news policy defines:

1. What the station considers to be news
2. How news will be reported
3. How the newscasts will be organized
4. How much emphasis will be placed on international, national, regional, and local news
5. How much emphasis will be placed on hard news and on feature news
6. How many people will be employed in the news department
7. What the responsibilities of the news staff will be

The first question to be answered in a news policy is: what is news? In any given community, each radio station may define news differently. One may put a great emphasis on local news. This station may cover local meetings such as the city council, zoning commission, city planning commission, and a wide variety of civic activities. This station may also cover local automobile accidents, fires, school closings, weather warnings, and any event that has some impact on the lives of citizens in the local community.

Another radio station in the same market may cover predominantly national and international news. This station may not consider automobile accidents, fires, and local zoning meetings to be of major significance to its listeners. Since one station in the market area places heavy emphasis on local news, this station may find its niche by providing complete national and international news.

The station that stresses local news may have ten short newscasts throughout the broadcast day, while the station that stresses national/international news may have just two newscasts, but each may be a half hour or longer in length. Both stations may offer some special news programs, but these also differ. The local news station may offer short feature news stories, while the national/international news station may offer longer programs from nationally known news commentators, or they may offer in-depth analyses of news events.

The news policy will dictate everything at your station from job responsibility to writing style. As we explore the various aspects of the reporter's job, we will see the significance of the station news policy.

News Beats

A "news beat" is a particular news source that a reporter regularly checks with to get the latest information. The news staff at the station that stresses local news will have news beats for its reporters—the police station, city hall, the mayor's office, the courts, and so forth. The reporter will probably establish a contact person at each beat—someone to contact every day to find out what is happening.

Most news beats are contacted via telephone. The reporter telephones to his or her news source and frequently records the source's information on tape for later broadcast.

In addition to telephone beats, the reporter at the local news station should be active in the community—attending meetings, visiting the site of local events, driving around the city in the station car.

Meanwhile, the news staff at the station stressing national/international news may not have local news beats. Their "beats" may be particular newspapers or magazines that they read on a regular basis that provide background information on national/international people and events. The news staff also develops research files that will provide material for analyses, background, and commentary.

The Audio Actuality

The local news station staff may place a great emphasis on getting actualities for each newscast. An *actuality* is a brief statement or comment in the voice of someone close to an event. It can be the mayor, a councilwoman, the police chief, an eyewitness to an event, or a local personality. Anyone who can add additional insight to a story is a good candidate for an audio actuality. The actuality is usually short—five to ten seconds. It gives the newscast vitality, immediacy, and realism. The actuality varies the pace of the news program and takes the listener to where the news is actually happening. Many stations that stress local news run several actualities in every newscast.

The station that specializes in national/international news may run no local actualities, but they may use audio feeds from Washington, D.C., from federal bureaus, or from national/international news services. The audio feed is generally sent over the telephone lines at specified times each day. The local station records the message to use for a later newscast. AP and UPI have audio feed services and many federal agencies provide free audio feeds to any station that wants to telephone for information. *The Wall Street Journal* provides an audio feed of financial news to stations that carry their advertisements. The national/international station may rely on these kinds of audio feeds rather than relying on local actualities.

News Styles

Delivery styles may also be different at the two stations. The local news station may present the news with more vitality and may have sound effects that open and close the newscast. The national/international news station may present the news in subdued style.

The news policy at a station should be in writing and available for the news staff and should indicate specific details of writing style, length, and word choice for news stories. The station news policy will dictate the scope and depth of the station's news coverage. A new employee needs to study the policy to understand fully the job of preparing a newscast. If the station does not have a news policy, one can be developed through conferences with station management. A news policy is a very important document and will provide a sense of purpose and organization for the news staff.

THE NEWS OPERATION

The News Staff

The news operation at a station will definitely be dependent upon the size of the news staff, which may range anywhere from one to fifteen or more.

A large station in a major market with a heavy news programming schedule may employ a large news staff. There will be a news director who is responsible for the entire news operation, and there will be several reporters, several air personalities, and some part-time reporters and part-time air personalities for evenings and weekends, and perhaps some stringers and interns.

A stringer is a person who is not on the regular payroll of the station but who is paid for the stories that he or she covers for the station—a free-lancer. Occasionally the station may ask the stringer to cover a special story. The stringer covers a particular location, frequently a suburban area outside a major city. Broadcast journalism students often serve as stringers for radio stations or wire services.

The small station may function with a two-person staff—a regular person for the weekday shifts and a part-timer for weekends. A very small station may not subscribe to a news wire service. In this case, the news staff clips articles out of the local and regional newspapers and prepares a news summary.

A news operation may range from the sophistication of WBBM, the all-news station in Chicago, to the one-room station in rural Iowa that broadcasts newspaper summaries. No matter what the size of the news operation, there is probably room for more news options — our next topic for consideration.

News Options

News programming does not have to be limited to the studio announcer who reads the news from wire service copy. There are many options available that will bring vitality to the newscast and will better serve the station audience.

The audio actuality has already been mentioned as one option that brings immediacy to the news. Some stations try to have two or three actualities in every newscast.

Audio feeds, which have been mentioned earlier, are also good additions to a newscast. If a station becomes affiliated with one of the major radio networks, that station will receive regular feeds from the network — ABC, CBS, NBC, Mutual, Turner Broadcasting, and others. The station will receive news features and news analyses by nationally known personalities.

Some news services provide regular audio feeds at a modest cost. The Black Radio Network is a minority news service that feeds news of blacks, Hispanics, and Indians to subscribing stations. Capitol Assignment sends stories on news events occurring in the nation's capitol to subscribing stations. Farm Focus provides coverage of the futures markets, gives grain and livestock reports from major livestock markets, and reports other news that affects farmers. Exhibit 5.1 indicates some of the news services that are available to stations that want to make their news coverage more than just an FCC requirement.

Other ingredients of the station sound include weather reports. These may be obtained by calling the local weather bureau and having a station announcer report the information. A station that wants to provide its listeners with more complete information may subscribe to a weather service for audio feeds. Excellent weather reporting may contribute to a unique station sound and may give the station a competitive margin in the community.

Some agencies provide free audio feeds to stations. The station need only pay for the telephone call. A list of some of these agencies is provided in Exhibit 5.2.

The best method of learning about the availability of audio feeds is to read regularly the professional broadcast publications. These include: *Broadcasting, Broadcast Management/Engineering, Variety, Billboard,* and the numerous publications from professional organizations — The Radio News Directors Association, the National Association of Broadcasters, the National Association of Educational Broadcasters, Sigma Delta Chi, and others.

ABC Contemporary News Service
ABC Direction Radio News Service
ABC Entertainment News Service
ABC News on the FM Network
ABC News on the Information Network
ABC Radio Enterprise News Service
ABC Radio Network News Service
ABC Rock News Service

All of these news services are affiliated with the ABC network. They are provided via satellite transmission. Each news service provides hard news, and some of the services provide features, special reports, and commentators.

CNN Radio

Half-hour news programs featuring a full range of topics such as health, national weather, sports, and human interest. Satellite transmission.

Commodity Focus

Sixty-second reports on the latest commerce news. Provided by United Press International.

Discovery

Short programs sent daily about new and interesting developments in science. Information for the program is provided by *Popular Science* magazine. These programs are provided free to stations from Cinema/Sound Ltd.

Dr. Loraine Stern: Pediatrician

Short informational programs about insights into raising babies and children. These programs are produced by Radio Works.

The Nation's Weather

Weather roundup four times daily. Produced by AP Radio Wire.

The Original Fish and Game Forecaster

Complete news and trends in fishing and game. Produced by Station Program Resources.

Exhibit 5.1.   News and News Feature Sources

Source: *The Radio Programs Source Book*, 2d ed. A publication of the Broadcast Information Bureau, Inc., 1983. These are just a few examples of the many news options available in radio.

The American Character

Ninety-second programs about contemporary Americans and their positive contributions to American life. Produced by Inform Broadcast Services.

(continued on page 106)

Exhibit 5.2.   Free Programming on News-Related Topics

Source: *The Radio Programs Source Book*, 2d ed. A publication of the Broadcast Information Bureau, Inc., 1983. These are just a few examples of the many free news options available in radio.

(Exhibit 5.2 continued)

Animal Album	Daily programs about caring for and raising all kinds of pets. Produced by David J. Clark Enterprises.
Audio Feature Feeds	Audio spots—sixty- and ninety-seconds—on a variety of topics including news clips, talk, and feature. Produced by Audio Features Inc.
Family Health	Short features answering many of the most commonly asked questions about health care. Produced by Ohio University Telecommunications Center.
Minding Your Business	Feature programs about today's business. Information from *Inc.* magazine. Produced by Norwood Productions.

Featuring Local Talent

Some stations feature a regular personality who comments on people and events in the news. Stations may have sports features or a local personality who comments on local events or nostalgic topics.

These short feature programs do not need to be a great financial burden to the station. They can be prepared by someone in the community or they can be written by a station staff member.

A five-minute feature program that is prepared once a week can be aired several times during the week. The program can provide the citizens with some local perspective on the news or some insight into everyday living.

A five-minute feature program can add an ingredient to your station's newscast that no other station in the market offers. Feature programs require a news director who is willing to experiment and who has the creativity to imagine news programming that is more than just rip 'n' read.

Exhibit 5.3 is an example of a news feature program that is presented weekly over WHO radio in Des Moines. Jack Shelley, native Iowan and retired broadcast journalist, comments on news and events in small communities throughout Iowa. This is just one example of the type of feature programming that can be used.

HOME TOWN NEWS COPY/JACK SHELLEY/7/15/78

How do you do again, everyone; this is Jack Shelley with the Home Town News.
Having spent the last year or two trying to persuade some wrens to take up residence in the nice little house we put up for them, I must admit to a certain sense of frustration when I look at the front page of a recent edition of the Sac Sun, at Sac City, Iowa.

(continued on page 107)

Exhibit 5.3.   Sample Feature News Programs

(Exhibit 5.3 continued)

"Mobile Home for Birds," the headline says. There are photographs to go with it too. And it's quite a story—the story as the Sun puts it, of "two wrens with a rather unusual attitude about where to build a home for their soon-to-be-born family."

Most birds are content to build their nests in trees or on a window ledge or maybe in a big bush. Birds hardly ever build their nests in the bumper of a car.

But that's what these wrens at Sac City did. The tiny, feathered Momma and Poppa flew to the bumper on a car . . . found a small opening just to one side of the spot where the license plate is fastened onto the front bumper—and chose the small, enclosed space they found as the place where they built their nest and hatched their eggs.

The Sun has pictures of one of the wrens flying into that little opening beside the license plate, and another close-up of two baby wrens' mouths, wide open in the usual appeal for Momma and Poppa to feed them—with these little mouths right beside the county number, at one edge of the license plate on the front bumper.

Talk about a strange place for birds to build a nest! Remember—all this while the car was being used in normal service. As Mrs. Jarvis would get in her car and drive someplace, she didn't even know for quite a while that she was transporting a nest, with eggs that were soon to become baby wrens.

But once those eggs were hatched, the Sun says, it didn't take long to figure out what was going on. And on July 5th, when the story was printed, this is the way things were. When the Jarvis car was parked at home, Momma and Poppa Wren were always near it. They'd perch on a tree near the car and spend all their time popping into that hole near the edge of the license plate on the car's bumper—feeding their babies and making sure they were all right.

But when someone in the Jarvis family had to use the car and would drive away, the mother bird would usually fly right along with the car for a block or so—then return to the tree at the Jarvis residence and wait until the car brought the babies back home again.

What happens from now on, the Sac City paper says, nobody knows for sure. For example, when the baby wrens get big enough to start flying out of the nest in the bumper—how will they know where to find their nest if Mrs. Jarvis has driven the car away for a few hours?

"They'll likely survive," the paper says. "And when they're grown-up wrens, they'll probably tell their children—"Listen, you think you've got it rough? Why when we were kids, we didn't even have a permanent home."

Jack Shelley, WHO, Des Moines, Iowa

HOME TOWN NEWS COPY/JACK SHELLEY/5/27/78

How do you do again, everyone; this is Jack Shelley with the Home Town News.

Well, it went right by me without being noticed. Did you pay any attention to the unusual nature of the date, on the first Saturday in this month?

"Uncle Bud," in his column in the Clarinda Journal, says Jo Kaserman brought him up to date—if you'll pardon the pun—about what happened on that Saturday.

It was May 6, 1978, you see. And if you wrote the date down the way people often do, it was what is sometimes called a "numerically ordinal date." Now what does that mean?

Well, you could write it 5–6–78—get it? Five, six, seven, eight—all in a row. And Uncle Bud says Jo told him that's the first time this has occurred in 12 years. What's more, it won't happen again for another 11 years. Then, on June 7, 1989, you can write the date this

(continued on page 108)

(Exhibit 5.3 continued)

way: 6-7-89. That, by the way, will be the last time in the 20th century that we'll have one of those ordinal dates.

Of course, when we get to January 1, 2-thousand and one, we can write that 1-1-01, if that makes you happy—and if you're still around, which we hope you are.

It'll take more than 30 years after that before the numbers are all lined up in order once again. That will be January 2nd, 2-thousand and thirty-four—which you can write 1-2-34. Next time around it'll be 2-3-45—if you last until 2-thousand and forty-five. That date being sixty-seven years from now. I'm not going to hold my breath. But I am going to set my sights . . . and have my pencil all sharpened for that next time around in this century—June 7th, 1989—6-7-89.

Thanks, Uncle Bud. Let's hope we both make it.

Jack Shelley, WHO, Des Moines, Iowa

SPECIAL ASSIGNMENTS: THE INTERVIEW

The news staff may have a variety of special assignments in any given week. These may include conducting an interview, covering a press conference, preparing a documentary program, or reporting on a disaster. Each of these assignments requires some special preparation.

The interview is a daily assignment at many stations. As a news reporter, you may be interviewing nationally known persons or persons from the local community. You may interview an eyewitness to a crime, or a victim, a political leader, an entertainment celebrity, a law enforcement person, or persons from any of hundreds of occupations. Interviews are of two types—the studio interview and the on-location interview. Many of the preparations that must be followed are similar for both types.

Preliminary Steps in the Interview Process

The first precaution for a broadcast interview is: don't be overly timid about requesting an interview. The person whom you want to interview may be busy, but don't be too timid to call back several times, if necessary, to schedule the interview. Always be polite and try to schedule the interview at the most convenient time for your guest.

When an interview is scheduled in advance, be sure you have prepared for the interview. Know your guest and know your topic. Some time should be spent in research. If you are interviewing an author, be certain that you have read some background information about the author and that you have read some of the author's works. If you are interviewing a resource person about an issue, know as much as possible about the issue so that you can ask meaningful questions. Careful research is vital to a successful interview.

Prepare a list of questions for the interview. Be prepared to divert from your list, but have the list of questions with you to be certain you cover the questions you believe are essential. Also, if you are nervous, a list of questions will give you confidence—you've done your research. You are prepared!

Whether the interview is in the studio or on location, an essential step is to check your equipment before the interview. It is very embarrassing to arrive at an interview site and have to learn how to use your equipment while the person you want to interview waits. The same idea applies to studio interviews. Be certain that batteries are charged, that microphones work, and that the recorder is not malfunctioning. In the studio, a board operator should be on duty to record the interview.

At the Interview

If the interview is being conducted on location, try to avoid locations where background noises will interfere with your recording. Some background noise can give your interview a sense of location, but when such noise interferes with your understanding of the guest or detracts from the information quality, the background noise is too loud. You want just enough noise to establish location but not enough to interfere with the interview.

In the actual interview, try to put your guest at ease. Many individuals are nervous about being recorded for a broadcast. Before beginning the interview, chat with your guest. Try to make your guest feel as comfortable as possible. Remember that you are the host.

When you begin the interview, listen to your guest. Begin by asking a question from your prepared list. Listen to the answer. Occasionally the answer will demand a follow-up question for which you will not be prepared unless you have listened carefully. Sometimes the answer to a question may lead in an entirely new and important direction, but if you are not listening to your guest, you cannot lead your guest down that new avenue.

Conducting an interview is hard work. You must ask your questions, listen to your guest, evaluate the answer, decide what question to ask next. Your mind is very active—interacting, listening, evaluating.

Preparing and Asking Questions

There are some standard rules governing interview questions, and these you should study before preparing your own interview questions.

Avoid loaded questions. A loaded question is one that places the guest in a bad light no matter what the answer. The classic loaded question generally used as an example is: When did you stop beating your wife? Guilt is assumed in the question. This type of question places the guest in a negative light no matter what the answer.

Avoid questions with yes or no answers. A question that can be answered with just a yes or a no will not advance your interview very far. If you are interviewing a local celebrity about a book on local history, you could ask: "Your great-grandfather was born here in Boone County, is that correct?" The guest may answer "yes," and the interview hasn't progressed very far. The guest could be asked: "Tell us when and how your family settled in Boone County." This gives the guest a specific direction and allows the guest to tell his or her own story.

Avoid ambiguous questions. If you are interviewing a law enforcement officer about a crime, you could ask: "What do you think about the circumstances in this case?" But the officer will not know what information you are seeking. The question is ambiguous. Ask specific questions: "What evidence do you have to connect the suspect with the crime?" "When was bail set for the suspect?" Without a specific direction, you will lose control of the interview.

Avoid two questions posed as one. If you are interviewing a political leader about a piece of social security legislation, you could ask: "What does the bill mean and what do you think about it?" The guest will probably have forgotten the second part of the question before getting to it. These are really two questions that could lead in two very different directions. You are likely to get clearer and more complete information if you pose major questions separately, not as a package joined by ands.

Avoid overly complex questions with repetitive clauses. "We know this legislation will introduce major changes in social security policy. Would you give us specific details of these changes and tell us about the nature of the changes inherent in the legislation?" The question is complex and repetitive. The question could have been simplified: "Would you please explain the changes that will be introduced in this new social security legislation?" Make your questions clear and specific. If you follow the rules for developing good questions, and if you prepare for the interview, your chances of getting a good interview are greatly increased.

Controlling the Interview

Finally, there is one major *do* for your interview. Do maintain control over your interview. Don't stifle the guest's answers, but try to keep the guest on target. When a guest rambles, you will have a great quantity of tape to edit later. If you allow the guest to ramble too far, you may have few specific answers to your questions. Follow up interesting answers, but keep the guest on the central questions.

If you have carefully researched your interview topic, have prepared a list of questions, and have checked your equipment, you are ready. In the interview, remember to maintain control, try to keep the guest focused, listen to the guest, and follow up on interesting answers. If you remember these things, you should have an interview that results in some good material for your news program.

SPECIAL ASSIGNMENTS: THE PRESS CONFERENCE

A broadcast journalist working for a radio station may be called on to cover a press conference. A press conference can be a large event for a nationally known person or it can be a local event covered by just a few reporters.

It is not uncommon for a political leader to call a press conference. The purpose of the conference could be to announce a candidacy or to make clear a point of view on an issue. The station may decide that the group calling the conference simply wants publicity and does not have any newsworthy information to share. In this case, the station may decide not to cover the conference, but the station's news director would be embarrassed to discover that the station missed an important news story by failing to cover the conference. If there is any doubt as to the newsworthy quality of the conference, the station will usually decide to send a reporter to the scene just to be certain that a good story isn't lost.

Be Prepared

The same rules apply to the preparation for the conference as for the interview. Be prepared. Conduct some research on the topic of the conference if time permits. Prepare a few questions so that you can pose a meaningful question that will make the press conference worth the time you spend attending. Be certain that your equipment works so you can take some good material back to the station.

Advantages and Disadvantages

The press conference has advantages and disadvantages for the broadcast journalist. The advantages are that the conference provides a forum for obtaining information. The disadvantages are: 1) you are not in control—the guest is in control, 2) you cannot ask in-depth questions; 3) rarely can you follow up on a question.

While the press conference is not the ideal environment for obtaining information, it does serve as an information source. The press conference can provide actuality material for your newscast, giving your regular newscast more vitality than just a rip 'n' read approach.

SPECIAL ASSIGNMENTS: THE DOCUMENTARY

The documentary format is an in-depth investigation of some event or issue. This event or issue can be serious—crime, military spending, community programs for senior citizens—or, it can be a lighter topic—sports, community recreation, a community pageant. The distinguishing feature between a documentary and any other news format is the in-depth coverage that a topic receives.

Documentaries and Minidocs

The main goal of a documentary is to give your listeners information on a topic from a wide range of viewpoints. If you are preparing a documentary on teen-age drinking in your community, you may want to interview a school counselor, the director of the youth shelter house, some students, some parents, and the police chief.

The goal is to provide the following types of information for your listeners: What is the scope of the teenage drinking problem in your community? Why do teens drink? What do teens think about drinking? How do counselors and public officials characterize the problem? What community services are available to teens with drinking problems?

There are two ways of presenting the information to your listeners. You can put all the information into one documentary—usually a half-hour—or you can cut the program into a series of minidocumentaries to air throughout the broadcast day or throughout a broadcast week.

Preparing the Documentary

Before you begin gathering the information for a documentary, you should allow plenty of time for research and thought. Read as much as you can about your topic; prepare a list of persons to be interviewed; prepare a list of questions for each person. When you begin a radio documentary, you usually do not know what you are going to find, but you should have a good idea what you are looking for.

After you have gathered all the materials, you will still need to produce the program. Production techniques are discussed in Chapter 6.

The documentary format is one of the most challenging and rewarding news forms for the broadcast journalist. Too few radio stations take the opportunity to serve the communities through this format.

SPECIAL ASSIGNMENTS: DISASTER NEWS

Your station should have a plan of action for disaster news that should be part of the news policy at the station. What kind of coverage will your station provide and how will the station provide its coverage? These are crucial questions that cannot be answered during a disaster.

If your station news policy states that you will be prepared to give your listeners complete and uninterrupted coverage during a disaster, you should have a file of information on the disasters that are likely to occur in your community. If, for example, your community is subject to hurricane threats, you should have a file of information on this topic. When did the last hurricane hit your area? How much damage was done? How does the national weather service track hurricanes? How many hurricanes hit the U.S. in recent years? What precautions should your listeners take? A large file of background information the announcer can draw on will aid the station in remaining on the air for long hours during the disaster.

The station news staff should decide whether mobile units will be used and how many reporters will be on duty. Every member of the news staff should know the station policy regarding disaster broadcasts. That way, if a disaster strikes on a weekend when only the part-timer is on duty, that person can take the necessary steps to begin the disaster coverage process.

JOURNALISTIC RESPONSIBILITIES

Accident Reports

Recently, the news media in Washington, D.C., made news by reporting the death of the mayor who was, in fact, not dead. Someone telephoned the stations and reported that the mayor had been shot. Without properly investigating, the stations reported the information to their listeners. The news personnel were very embarrassed when the mayor emerged from a meeting to declare himself alive and well. The mayor also rushed to telephone his family, who would have been shocked had they heard this false news.

This true story illustrates the need for careful investigation of all news items. It is particularly important to be careful if you are reporting the death or injury of someone in your community. A false story may cause great misery and may subject your station to lawsuits for the damage and misery you have caused.

Reports of automobile accidents, fires, and other emergencies call for extra caution. At the scene of a fire, rumors can spread almost as fast as the flames. Some rumors may indicate no deaths, while others may indicate multiple deaths. The reporter should carefully check the fire or police reports before releasing the information to listeners. The same precaution should be practiced at the scene of an automobile accident. Never release the names of accident victims unless you can verify that indeed you have the person's right name and that you are reporting the person's actual condition. Police reports are excellent sources for verification; also, a call to the local hospital can be used as a double check. Though speed is important in radio journalism, accuracy is even more important!

Objectivity and Fairness

Objectivity in news means providing your listeners with unbiased information. Fairness means providing complete coverage of news events in a balanced manner.

Fairness and objectivity frequently need to be considered. Let's say, for example, that there are two spokespersons for a controversial issue. One person is very dynamic and articulate; the other is a poor spokesperson and does not well represent the other point of view. Do you seek another spokesperson who can better represent the side that will be poorly presented?

For another example, let's say there are two articulate spokespersons for a very controversial issue. Do you give each person a chance to represent his or her view, or do you locate a more neutral person who can represent both sides?

Decisions about objectivity and fairness are often difficult. As a broadcast journalist you need to strive for objectivity and balance. You need to be able to step back from a story and ask yourself if you are being fair. Is there a better way to provide the information to your listeners? Have you presented a balanced view? Is the information that you are presenting actually representative of the situation? By frequently questioning yourself about fairness and objectivity, you are more likely to do a better job of providing news for your listeners. You will never achieve an absolutely fair and objective report, but you should try for maximum fairness and objectivity. Remember that, in a democracy, citizens base their decision making on the facts they have learned. As a broadcast journalist, you are an essential ingredient in the fact-finding process for your listeners.

IN SUMMARY

This chapter has provided some basic information about radio journalism. If you want to make radio your career, you will need to gain additional information. The bibliography to this chapter provides some sources for additional reading.

One of the major services that radio provides to a community is news, and Americans have learned to rely on radio as their most immediate news source.

BIBLIOGRAPHY

Bittner, John R., and Bittner, Denise A. *Radio Journalism.* Englewood Cliffs, N.J.: Prentice-Hall, 1977.

Brooks, Brian S.; Kennedy, George; Moen, Daryl R.; and Ranly, Don. *News Reporting and Writing.* New York: St. Martin's Press, 1980.

Callihan, E. L. *Grammar for Journalists.* Rev. ed. Radnor, Pa.: Chilton, 1975.

Garvey, Daniel E., and Rivers, William L. *Newswriting for the Electronic Media.* Belmont, Calif.: Wadsworth, 1982.

Hohenberg, John. *The Professional Journalist.* New York: Holt, 1969.

Johnson, Joseph S., and Jones, Kenneth S. *Modern Radio Station Practices.* Chapter 8, "News." Belmont, Calif.: Wadsworth, 1978.

Mencher, Melvin. *Basic News Writing.* Dubuque, Iowa: William C. Brown, 1983.

Shook, Frederick. *The Process of Electronic News Gathering.* Englewood, Col.: Morton, 1982.

White, Ted; Meppen, Adrian; and Young, Steve. *Broadcast News Writing, Reporting, and Production.* New York: Macmillan, 1984.

QUESTIONS FOR REVIEW AND DISCUSSION

1. Why does radio have a distinct news advantage over every other medium?
2. When do people rely on radio news more than on any other news source?
3. How should news from the wire services be organized?
4. What should be included in a station's news policy?
5. What news options are available for a regular newscast?
6. What is a stringer?
7. What is an audio actuality?
8. What are some sources for audio actualities?
9. What sources are available for audio feeds?
10. How can a broadcast journalist learn about new audio feed sources?
11. How can local talent be used in a station's total news programming?
12. What kinds of feature programming can a station offer?
13. How should a reporter prepare for an interview?
14. What kinds of questions should be avoided in an interview?
15. What should the reporter do during the interview?
16. What is a follow-up question?
17. What is meant by maintaining control over an interview? Why should the reporter maintain this control?
18. What are the advantages and disadvantages of a press conference?
19. What is the major difference between a documentary and other news formats?
20. What are the initial steps in preparing a documentary?
21. If you are preparing a documentary on a proposed property tax increase, what sources of information would you use?
22. Why should the station have a policy prepared for disaster broadcasts?
23. Why is accuracy so important in news coverage?
24. Give an example of a situation where accuracy may be difficult to achieve.
25. How can a reporter double check information sources?
26. What is meant by fairness and objectivity in covering the news?
27. Why are fairness and objectivity so important?
28. Provide several situations that present fairness and objectivity problems.

Chapter Six ---

RADIO PROGRAMMING

I wonder if I could have survived my childhood without the escape from it offered by radio. A fat kid, I could play on the same team as Jack Armstrong and still be reassured by the Fat Man that 'heavy' guys could be heroes in their own right. A bookworm, I could be smugly satisfied that avid readers such as Ellery Queen, Sherlock Holmes, and I really won out in the end. And, in the midst of poverty, Jack, Doc, and Reggie assured me that the only good thing to do with money was to get rid of it fast so you could get on to something important. . . . Each of us found his own escape, his own reassurances in radio . . . . As long as our generation lasts, radio will be one of the most imperishable art forms.

> Jim Harmon,
> "What Did You Listen to
> Every Day at Five O'clock?"
> In Kirschner and Kirschner,
> *Radio and Television,* 1977

Radio production is very challenging because radio is frequently a one-person operation. Since there is only one person in charge, the radio operator cannot rely on others to get the job done, nor can mistakes be blamed on others. This aspect of radio production that makes the job difficult also makes the job rewarding and challenging. What are the tasks of the radio producer and how can these tasks be fulfilled in the best ways? To answer these questions, this chapter explores radio production from the job of the board operator to the job of the radio producer.

THE ESSENTIAL ELEMENT: THE RADIO ANNOUNCER

The Station Voice

In radio the station voice is an all-important consideration, since the listener identifies the station announcer with the station itself. Rarely does the station announcer announce only. The announcer may be the music director who selects music for the station, or the news and public affairs producer, or even the station manager. A pleasing voice and the ability to use that voice under a variety of circumstances are just two assets of the station announcer.

What constitutes a good voice? A good broadcast voice is one that is clear, has good diction, is easy to listen to, and conveys a desire to communicate.

One major requirement for a good radio announcer is a voice that has no major speech impediments. A speech impediment is any vocal characteristic that interferes with communication. An impediment may be as obvious as a stuttering problem or may be an annoying speech pattern of which the individual is not aware.

For anyone interested in radio announcing, vocal problems such as a nasal voice or difficulty in changing the rate and the character of the voice should be eliminated. On most university campuses, help is usually available in voice and diction classes and in speech clinics.

The announcer must develop an on-the-air style that is authoritative, yet friendly and relaxed. The announcer must also project confidence so that the listener will have confidence in the station and in the products the station advertises. Listeners are embarrassed for the radio announcer who stumbles over words and is obviously unsure and uncomfortable. Confidence for the radio announcer must show itself as the announcer's ability to do the job well, not as arrogance or cockiness.

Different program formats may call for different vocal styles. A call-in program where local citizens exchange information may call for a friendly neighbor-to-neighbor approach, while a political call-in program may demand a crisp, no-nonsense approach for those callers who get out of line. A classical music station with no commercials may call for a professional style that doesn't interfere with the flow of the music, while a rock station may call for an announcer who gets directly involved in the music. The radio announcer must be able to adapt an announcing style to the demands of the station and of the particular program. Style, however, should never get in the way of good communication.

The Professional Announcer

One of the most important assets of a radio announcer is the ability to pronounce every word properly—even foreign names. The radio listener will frequently adopt a pronunciation used by the radio announcer, which means that the listener could be placed in an embarrassing situation unless the announcer has provided the correct pronunciation.

When the press services (AP and UPI) send stories over the wires, they frequently provide a pronunciation guide for difficult names and places. The guide divides the word into syllables and anglicizes the foreign words. The major broadcast networks—ABC, CBS, and NBC—publish pronunciation guides for their announcers and those at affiliated stations. In addition, the dictionary is always a good authority on pronunciation. The radio announcer should purchase some good pronunciation guides and should try to stay aware of new words and of the proper pronunciation of the names of world leaders.

Musical composers and their works are frequently challenges for the announcer. Difficulty with musical names and terms can be partially alleviated by taking courses in music appreciation in which students are exposed to a wide variety of musical terms, composer names, and musical compositions. Also, many courses in radio and television speech include a unit on musical pronunciation.

The most important element of good broadcast announcing has been saved for a final consideration. That element is a desire to communicate. No one wants to listen to someone who doesn't want to communicate. The desire to communicate with the listener is what keeps the audience loyal to the station. The broadcast announcer must develop a frame of mind that produces warmth, friendliness, confidence, authority, and pleasantness when before a mike. Further, the desire to communicate will make minor speech problems oblivious to the listener. Conveying an eagerness to speak directly with the listener is the most crucial element of good announcing.

Good announcers are not natural "talents"; they are highly trained professionals who have prepared themselves for a very difficult job. The essential element of any radio station is its "station voice," and the serious broadcasting student will take every opportunity to develop a wide variety of announcing and performing skills.

THE MUSIC PROGRAM

Operating the Board

The music program is the standard type of program at most stations where the live "announce" or disc jockey announces the musical selections and interacts with the listeners. Many stations are now using an automated system for much of their musical fare, but even if a station has an automated system, every radio operator needs to know how to get a program on the air manually. Also, even with automated systems, the radio operator is frequently called upon to introduce the music or read live announcements.

The music program that will be considered here is the program with the DJ personality. The pieces of equipment the DJ will probably use are a mixer, two turntables, a cart machine, and a reel-to-reel recorder. A five-minute segment from a music program may include the following:

1.	Theme music: on cart	15 secs.
2.	Opening announce: live (read by announcer at console)	30 secs.
3.	Record #1: turntable #1	95 secs.
4.	Record #2: turntable #2	85 secs.
5.	Advertisement #1: live (read by announcer at console)	15 secs.
6.	Advertisement #2: reel-to-reel recorder	30 secs.
7.	Record #3: turntable #1	30 secs.

Total time   5 minutes

While this is a relatively simple program, it still requires preparation before and during the broadcast. Before the broadcast, the announcer/board operator/DJ has a list of tasks to accomplish.

1. At this station, the cart machine with the theme music and the reel-to-reel recorder with advertisement #2 must be patched through the console. The board operator patches these two pieces of equipment through the console at the patching panel.

2. The board operator labels each pot to be certain that he or she knows where the sound sources are located on the board. The radio operator can be placed in an embarrassing situation when he thinks he is potting up on a commercial and pots up music instead.

3. The operator places the cartridge in the cart machine and sets a level on the board. Once a good level has been set, the cart should be allowed to recue. The level may remain set on the pot control, but the pot should be taken out of channel.

4. For all of the live announcing, the operator must set a level. While sitting at the console, the board operator reads the opening announce at the level he will use on the air. The announce pot is set and taken out of channel.

5. For the second advertisement, the reel-to-reel recorder is threaded and placed in the playback mode. The board is set for the proper level and the pot is taken out of channel.

6. Records are placed on turntables #1 and #2. The records are cued and proper levels are set. The pots are left in the out-of-channel position. (Exhibit 6.1 explains record cueing.)

After completing this six-step procedure, the DJ/board operator is ready to air the music program. But there is still much to remember if the program is to be aired free of errors.

1. Turn on the power switch but do not activate the speed control. (The speed control rotates the platter at $33\frac{1}{3}$, 45, or 78 rotations per second.)

2. The record is placed on the turntable with the needle placed at the beginning of the cut to be cued.

3. With your fingers on the edge of the record (never on the grooves), the record is manually rotated forward until the first sounds are heard. These first sounds will be rumbling noises, like the sounds of a deep groan.

4. When this noise is heard, stop the record and then rotate the record backwards until the noise is no longer audible.

5. If the turntable is an older model, the operator may need to rotate the platter one-half of a rotation (or more) back from the beginning of the first sounds.

Caution: The board operator does not want the record cued too tightly (too close to the beginning sounds), since this would cause what is called a "wow" sound as the turntable motor builds up to the proper speed. The record should not be cued too loosely (too far from the beginning of the sounds), since a loose cue would render dead air — silence when there should be music.

Exhibit 6.1.   Cueing Instructions for a Record

---

1. Theme music on cart (15 seconds)

   15 seconds into program
2. Opening announce (30 secs.)

   45 seconds into program
3. Record #1 (95 secs.)

   2 minutes and 20 seconds into the program
4. Record #2 (85 secs.)

   3 minutes and 45 seconds into the program
5. Advertisement #1 (15 seconds)

   4 minutes into program
6. Advertisement #2 (30 secs.)

   4 minutes and 30 seconds into program
7. Record #3 (30 secs.)

At 5 minutes into the program, fade the program out.

At airtime, open the air channel on the pot and punch the start button on the cart machine. At 13 seconds into the theme, begin to fade the theme music out.

At 15 seconds into the program, open the announce pot channel and deliver the opening announce.

Close the announce pot.

At 45 seconds into the program, punch the on button for turntable #1 and open the turntable pot.

At 2 minutes and 20 seconds into the program, close the pot for turntable #1 and punch the on button for turntable #2. Open the pot for turntable #2. While record #2 is on the air, cue record #3 on turntable #1.

At 3 minutes and 45 seconds into the program, close the pot for turntable #2 and open the announce pot.

At 4 minutes into the program, close the announce pot. Punch the reel-to-reel recorder on and open the recorder pot.

At 4 minutes and 30 seconds into the program, punch the on button on turntable #1, close the recorder pot, and open the pot for turntable #1.

SHOW TITLE __3:00 P.M. News__   AIR TIME __3:00 P.M.__

TIME	SEGMENT CONTENT	SEG TIME IN	SEG TIME OUT	Air Time In		Air Time Out
10 secs	Opening Theme	00:00	00:10	3:00:00		3:00:10
20 secs	News Item #1	00:10	00:30	3:00:10		3:00:30
20 secs	News Item #2	00:30	00:50	3:00:30		3:00:50
10 secs	News Item #3	00:50	1:00	3:00:50		3:01:00
30 secs	Advertisement #1	1:00	1:30	3:01:00		3:01:30

Exhibit 6.2.   A Sample Time-Segment Sheet for Radio Production

The above description of the jobs of the DJ/board operator indicates the quickness and accuracy with which the board operator must function. The successful board operator is always busy with work to be done in preparation for the next segment. At times it may even seem as if the successful board operator could use more than one set of hands. If the station is automated, the job becomes much easier, but the board operator still has much to do. Cartridges must be loaded into the automatic-play cartridge machines, tapes must be placed on the recorders, and usually live announcements are used in the program—weather reports, station identifications, and news reports.

To help plan the music program, the station should provide some time-segment sheets for listing the various segments of the program. Exhibit 6.2 shows a sample time-segment sheet. Since radio is a precise medium, times must be scheduled very exactly. The time-segment sheet should initially be done in pencil to allow for adjusting the segments to fit the total program time.

Categorizing and Filing Music

Before the DJ broadcasts a music program, he or she must select the music that will be used. Most stations have a substantial record collection. The largest stations in the country (the top one-quarter) get records free from the record companies. These companies want their records to be successful, and supplying copies of records to the stations in the large markets can help insure success for the records. The small stations generally play a recording after it has become successful in major markets.

For a small fee, the small to middle market stations may subscribe to a recording company, which entitles the station to receive the latest music from the company. The very small stations may be limited to the music they can afford to purchase outright from a tight budget.

No matter how a station receives its recorded music, the selections must be categorized and properly filed if the music is to be easily located. All stations have a music library, a room devoted to record storage. A card, placed in a card file, identifies the type of music, the artist, the selections on an album (with the length of time of each selection), the recording company, the date of the recording release, and the shelf location where the music may be found. Each record is checked in and checked out just as a book would be at a library. Larger stations frequently employ one or more music librarians who have the responsibility for labeling each new musical acquisition, preparing the card to be filed in the music card file, and storing the record in the proper location. Exhibit 6.3 on pages 124 and 125 shows three sample music library cards.

Even if a station is too small to afford a music librarian, the DJ and other station employees should spend some time developing a system for categorizing and shelving all music. A large music collection is of little use if the music cannot be easily located.

The DJ selects the music for the program with a particular theme in mind or with a particular type of music in mind. The station manager has already selected a format and may have dictated that certain types of music be used at certain times of the day. A rock station, for example, may switch to folk rock during the daytime hours when students are in school and housewives comprise the largest share of the audience, changing to hard rock at 3:00 P.M. when the schools are dismissed.

Music selections must fit into program time limits. The traffic department must be consulted to determine what advertisements and announcements must be aired at what times. With this information, the music may be planned around the advertisements and the announcements. After the schedule is developed, it is recorded on the time-segment sheet.

Distinguishing Your Station

One critic of American radio has argued that when turning to the various commercial stations in any location, all of the stations sound alike. The DJ has control of three factors that can make his or her program unique—*announcer personality, musical selections,* and *special programs featured.*

Two stations with very similar formats can sound very different just because of the personality of the on-the-air talent. Two classical music stations, for example, can have very different sounds. At one station, the announcer can merely "announce" the musical selections. At the other classical music station, the announcer can get directly involved with the music and the audience, providing information about the artist and background about the music and requesting listener response. Though the two stations have very similar formats, they will sound very different.

A second method for distinguishing one station from another is in musical selections. Even if a station has a standard format, one or two programs each day can offer the listener something that is unique. The choices are many. A station can have thirty minutes each day devoted to music for children, or an hour each day of "show tunes" from the theatre and the cinema, or "late night music," or a special jazz program, or ethnic, synthetic, or religious music. One MOR station does not have to sound like the other MOR station across town or the one in the next town. The same is true for rock stations or stations with any other format.

The third technique the DJ can use to distinguish his or her program is to use special program features. Local guests or visitors to the city may be invited to the program for brief chats. Special kinds of information may be incorporated into the program such as congratulatory announcements to new parents, menus for school lunch programs, announcements concerning recreational activities or community events, or special tributes to senior citizens or outstanding high school students. In the process of serving the public, the radio personality can be creative and can provide information that the public would otherwise not have.

Exhibit 6.3.   A Music Library Filing System
(below) Here are three cards. The first card is filed by composer, the second card is filed by musical work, and the third card lists the library's holdings for a particular artist. With this system, anyone looking for a particular piece of music can find it in one of three ways. *(opposite)* The photograph with the WOI employees in front of the video screen demonstrates a new filing system that some stations have recently adopted. The filing is done on computer. Anyone searching for a particular piece of music types in the name of the composer or the name of the musical work and the computer will indicate the location of the work in the library.

```
Beethoven                                        LD 2105
                                                 File No.
Symphony No.9 in D minor, Opus 125 "Choral"

          Talent Yeend,Lipton,Lloyd,Harrell,
          Westminster Choir
          New York Philharmonic, con.by B.Walter

          Columbia ML-5200

          65:50

                                      WOI RECORD LIBRARY
```

```
          Suite No. 3 in C, for unacc. viola

                                                 File No.
          Bach

                    Talent Fuchs, viola           LC 2006




                                      WOI RECORD LIBRARY
```

```
                                                      # 7
BERNSTEIN, Leonard (conductor)

          SD 2953      SD 2972   SB 3139   SD 3024

          SD 2954      SA 1855   SD 3041   SB 3190

          SD 2955      SD 2993   SD 3020   SA 1862

          SD 2956      SD 3018   SD 3021   SD 3098

          SD 3105      SD 3062   SD 3022   SD 3101

          ST 23X       SD 3082   SD 3023   SD 3102
                    WOI Record Library
```

PRODUCTION ASSIGNMENTS

Tape Editing

Editing is one of the most creative aspects of radio. The editor can decide what information will remain in a program and what information will never be disclosed to the listener. A tape editor can make a faltering speaker sound fluent or can exacerbate distracting speech mannerisms with the careful splicing of the tape. The broadcast editor has a tremendous responsibility to be fair to the individuals who were interviewed and to represent accurately the content and the nature of an interview. The essential materials for editing are a sharp single-edged razor blade, splicing tape, and an editing block with a diagonal slit on which to cut the tape. An editing block is usually provided on every control room recorder. Exhibit 6.4 provides detailed instructions for editing. Exhibit 6.5 on pages 128 and 129 demonstrates the editing process.

1. First, dub the master tape (make a taped copy). Do not cut the original tape, since an editing error may damage the master. If a major error occurs while editing the dubbed tape, the master tape is still available and the program can be salvaged.

2. Place the dubbed tape on a reel-to-reel recorder.

3. Listen to the part of the tape to be edited. Decide exactly what is to be edited out. On a piece of paper, write the first and last words to be edited out of the tape. Listen to the tape again to be certain that the parts to be edited out of the tape have been correctly identified.

4. When deciding where to make an edit, remember that editing is easiest when there is some pause between words. Editing is difficult when one word runs into another. A good editing point occurs when an individual pauses momentarily. The criterion for editing, however, must be based on informational content, not on ease of editing.

5. Play the recording, listening carefully to the last word to remain on the tape. Stop the recorder as the end of the last word passes over the playback head. (This is just before the first word to be edited from the tape.)

6. Rock the reels back and forth with your hands to be certain that you have located the exact spot to be edited. You will hear a deep groaning sound as the last word rolls across the playback head and a whishing sound (the rushing sound of air) as a momentary pause occurs. Continue to rock the reels forward and you will hear another groaning sound as the next word begins. It is between the two words, in the pause space, where the edit should be made. If two words are uttered together, the editor must listen carefully for the very slight momentary pause or for a slight change in frequency caused by the formation of the speaker's new word. Editing between words where there is no pause can be extremely difficult. With practice, however, a good editor can hear the change from one word to another by slowly rocking the tape reels back and forth.

7. The exact spot to be edited should be located immediately past the playback head. A small mark should be placed on the nonmagnetic side of the tape with a felt tip pen or a grease pencil. Be careful that the mark is made on the tape, *not on the recorder head.* Ink or grease marks on the tape recorder heads will result in poor quality recordings. *Edit marks should always be made only on the tape.*

8. Now push the start button and play the tape to the last word to be cut from this segment. Again rock the reels to be certain that this is the exact location for the cut. Mark the tape with a felt tip pen or a grease pencil just past the playback head.

9. Rewind the tape to the first penciled mark. Remove the tape from the gates and cut the tape on the diagonal slit in the editing block.

10. Go forward until you find the second mark on the tape. Push stop, and cut the tape on the editing block.

11. Match the two diagonal cuts so that the two pieces of tape fit together snugly. There should be no air gaps. Now place a piece of editing tape about an inch long on the nonmagnetic side of the tape. Rub the tape with your fingernail or the capped end of a ball point pen until air bubbles have been removed from the tape and the two diagonal pieces of tape adhere.

12. Play the edited portion of the tape to be certain that the edit is good. A good edit is not noticeable to the average listener; the information flows smoothly and there is no loss of technical quality. In the process of editing, never discard a piece of cut tape until you are certain that the edit is exactly what you wanted. A good idea is to place the cut tape across your lap exactly as it comes off the machine. If you make a mistake in editing, the tape can be spliced back onto the reel with no damage done. If the tape has been discarded into the trash can, however, you may never find the piece you need among the mounds of tape that have already been tossed into the trash. If the edit is a good one, you are ready to move on to the next segment to be edited.

Exhibit 6.4.   Instructions for Editing an Audio Tape

Production Organization

When doing any kind of major radio production, the most important element is organization. The producer must have a clear idea of the necessary steps in the production process. Making a list in priority order is the best way of insuring that all the steps are accomplished on time.

The following list demonstrates the kinds of production activities that the radio producer must accomplish. The program being planned is a major radio documentary.

1. *Develop the major program idea.*
2. *Research the major idea* in the library, in newspapers, in journals, in magazines.
3. After the research, *prepare an initial program outline with specific subheads.*
4. *Make a list of persons to be interviewed.*
5. *Schedule the interviews.* Plan on conducting the most important interviews first.
6. *Prepare for each interview* with a list of questions to ask each person.
7. *Check all equipment before your interviews* to be certain the equipment is working correctly and that you understand how to operate the equipment.
8. *After the interview, label and file each tape.* It is also a good idea to note recorder meter readings for specific questions or for specific information. This will save a tremendous amount of time in the editing process.
9. After the interviews are complete, *prepare a final program outline with the specific information to be used in the program.*
10. *Schedule studio time for editing.*
11. *Write the open, the close, and the continuity between interview segments.*
12. *Edit the program.*
13. *Critique the program.*
14. *Do a final edit and a final dub.*

A production list is essential for any type of major radio production. For a radio drama, the list would include some different items—select actors, schedule rehearsals—but the production checklist would serve as the guide for accomplishing the production tasks.

Recording on Location

For many types of radio production, location recording is essential. When recording on location, there are a number of points to remember to insure a good recording.

First, think carefully about all of the equipment that will be needed. Many interviews have been lost because a cable is missing or because the micro-

(a)

Exhibit 6.5.   Editing Audio Tape

(a) Mark the tape with a felt-tip pen or a grease pencil just past the last word to remain in the tape. The mark should be made on the tape, *never on the playback head.*

(b) Roll the tape forward and mark the tape just after the last word to be edited from the tape. (The tape can be rotated by hand if the last word is close to the first word or the "play" button can be used for long pauses between the first and the last word.)

(c) Rewind the tape to the first mark. Remove the tape from the gate and place it on the editing block. Cut the tape on the diagonal. Roll the tape forward to the second mark. Remove the tape from the gate and place it on the editing block. Cut the tape on the diagonal.

(d) Shove the two pieces of cut tape together snugly so that the diagonal cuts match with no air space between them. Place a one-inch strip of editing tape across the edit. Rub the tape to remove air bubbles and to insure a tight bond between the two strips of tape.

---

phone is not in the side pouch of the recorder. An equipment list is frequently helpful in remembering the essential equipment.

Check all equipment before leaving the studio. Many interviews have been lost because a battery was not charged or a microphone jack did not match the microphone socket in the recorder.

At the recording location, check the recording site for extraneous noise. Loud traffic noises or loud noise from machinery may distract from the interview. Try to find the best recording location.

For some locations, the background noises add realism and flavor to the production. Interviews with tavern patrons concerning a presidential election wouldn't be appropriate without the sounds of people talking and glasses clacking in the background. An interview with a child on a playground should have the sounds of other children in the background. In these instances, the producer would establish the setting for the interview in the introduction so the listener is prepared for the background sounds. It is important that the background noises should not be so great as to distract from the information in the interview itself.

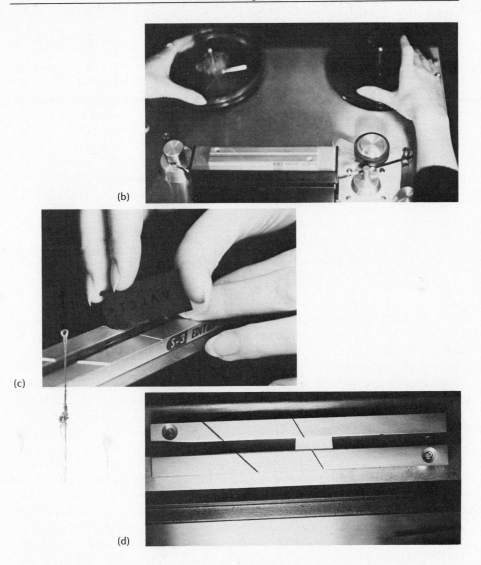

(b)

(c)

(d)

Monitoring the habits of the person being interviewed is also essential for a good recording. The individual may rock back and forth in a swivel chair, making her voice fade into and out of the microphone range. Or the individual may bang a hand on the desk next to the microphone each time he makes a point. If this occurs, some of the best information in the interview may be lost. It is not rude to remind the individual that these sounds are being picked up by the microphone. After all, the individual being interviewed is just as interested in the quality of the interview as you are. If you fail to detect distracting habits of the person being interviewed, the final product may be lacking in professional quality.

Using Music and Sound Effects

In some types of radio production you may want to use music or sound effects. A sound effect should never be inserted into a program without careful consideration. Before using music or sound effects, the radio producer should ask the following questions:

1. Does the music or sound effect add life or vitality to the program?
2. Does the music or sound effect move the story or the narrative along?
3. Does the music or sound effect fit the mood of the program?
4. Does the music or sound effect fit the time period and the environment of the program?
5. Does the music or sound effect blend into the program or does it disrupt the flow of the program and distract from the program?

Sound effects should never be used unless they are needed! Sound effects and musical bridges should blend into a program and should not stand out as separate from the program.

Sound effects libraries (a series of sound effects on records) are available and relatively inexpensive. Sound effects can be located for just about any effect you may want to create. If a sound effect is not available on prerecorded records or tapes, you can create your own. Galloping horses can be created by thumping your fingers on a table near a microphone; a roaring fire can be created by crinkling a piece of cellophane near a microphone. Creating sound effects can be one of the most challenging aspects of radio production.

When using music and sound effects in a radio production, the script must be carefully marked. If an effect is to be heard at a precise moment, circle the word on the script and indicate in the margin that the effect should be heard as this word is uttered. The volume level for the sound effect should also be indicated, and the exact point at which the music or the effect should be faded out or faded under should also be noted.

Using music and sound effects in audio productions requires some special production procedures. But with careful planning and with good production techniques, the results may be well worth the effort.

Assembly Methods

When recording a radio drama or radio feature program, one of two program assembly methods can be used. If recording a drama, the talent can be recorded in one session, the tape edited in a second session, and music and sound effects added in a third session. One major advantage of this type of program assembly is that the audio producer can concentrate on *one* aspect of the program in each recording session. This means that the talent can get good coaching from the producer; later the producer can devote full attention to sound effects and music. One major disadvantage is that this assembly method is very time consuming.

The other procedure that can be used is to record talent, music, and sound effects at the same recording session. This assembly technique is usually accomplished with a multitrack tape recorder and with several people working in the control room.

Many studios have a four-track recorder, which can be very useful in dramatic productions. The four-track recorder allows four independent recordings to be made at the same time. The talent can be recorded on one track, the music on a second track, sound effects on the third track, and the announcer can be recorded on the fourth track. All of the tracks can be recorded at the same session. The tracks can then be played back through the console, and music and sound effects can be adjusted to the proper level. The complete program can then be dubbed on a full-track recorder with the four tracks mixed into one. (See Exhibit 6.6)

This production procedure has advantages and disadvantages. If talent, music, and sound effects can be recorded at one session, this procedure requires a fraction of the time required for recording talent, music, and sound effects separately.

By using a multitrack recorder in a master recording session, each track is protected from mistakes that may be made on another track. If a sound effect comes in at the wrong time, that effect can be easily rerecorded. When such mistakes in the recording can be corrected, time saving is still realized on the portions of the master recording that went smoothly.

Another advantage is that the talent gets a sense of the complete production and can take cues from the sound effects and adjust interpretations to the music and sound effects. Under these recording conditions, the actors can render a more realistic performance.

A major disadvantage is that this assembly procedure is extremely difficult.

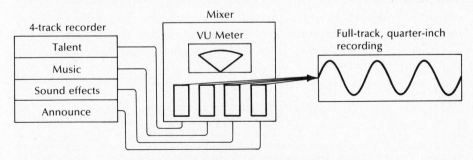

The multitrack recorder allows several tracks
of sound to be recorded independently and
mixed through the audio console into a
final blend of one signal.

Exhibit 6.6.   Multitrack Recording Procedures

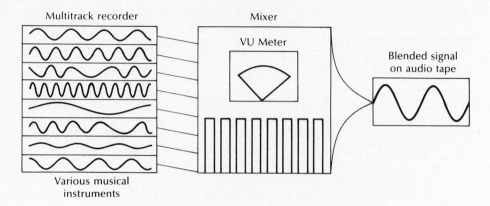

When recording various musical instruments on a multitrack
recorder, each instrument can be recorded on a separate
channel and mixed through the audio console. The final,
blended signal can be recorded on a monaural or stereo
recorder.

Exhibit 6.7.   Music Recordings Using a Multitrack Recorder

Working with talent, bringing in sound effects and music at the proper
moments are all difficult assignments. To accomplish all of these tasks requires
tremendous organization and assistants who are very familiar with the script.

Multitrack recording is the method almost always used when mixing music.
With a multitrack recording, individual musical instruments can be recorded
on separate tracks of the recorder, and the multitrack recordings can be mixed
through the audio console to provide just the right blend of musical instruments. (See Exhibit 6.7.)

Multitrack recorders can now record up to sixty-four tracks of sound. Each
track is an independent recording, giving the producer complete control of the
final sound. Using multitrack recorders is, in some ways, no more difficult than
using a single-track recorder—each track is treated as a separate recording
machine. But the final mix of a multitrack recording is both an art and a
science, requiring an understanding of the equipment and a good ear for mixing
the most pleasing sound.

THE AUDITION TAPE FOR SELLING YOURSELF

Any student interested in a career in radio broadcasting should know how to
prepare an audition tape. This tape is designed to demonstrate the student's
abilities in broadcasting. The audition tape is comparable to the artist's port-
folio. If the audition tape is to sell the student to the station manager, it should
be a sample of the best work the student has to offer.

The audition tape should be short—between five and ten minutes in length. If a job opening at a radio station has been announced in a variety of industry publications, the station manager may receive hundreds of audition tapes and will not have the time nor the patience to listen to lengthy works.

The sample tape should contain a variety of skills. Generally, the station manager is looking for someone who is flexible. The tape may include a segment from a music program on which you were the DJ, a segment from a news program that you wrote and delivered on the air, a segment from an advertisement you produced, segments from documentaries, dramas, or instructional programs that you prepared. Remember, the station manager is interested in listening to *short segments* of your work, not to lengthy passages.

The important point is to demonstrate that you can do more than announce music and rip 'n' read news. You may want to prepare more than one audition tape. You may want one tape for public broadcasting stations and another for commercial stations. For the public broadcasting station, you may want to demonstrate your skills at announcing classical music, delivering the news, interviewing guests, producing documentaries or dramas. For the commercial stations you may want to demonstrate your expertise at introducing popular music, interacting with the audience, producing and delivering news, producing public affairs programs, and producing advertisements. Short samples of each of these will suffice to demonstrate your skills.

On the audition tape, you want to reveal your personality. You may want to begin your tape with a brief introduction to yourself and explain to the station manager what will be heard on the tape. You should make clear your role in producing each segment on the tape. If your tape includes a segment from a radio drama, for example, don't just announce that a segment from a radio drama is on the tape; rather, explain that you edited the drama and mixed the music and sound effects or that you were one of the actors, or that you cut the script and served as assistant producer. The station manager wants to know the exact nature of your training, and by explaining your role in each segment, you clarify your skills.

The tape should be technically perfect. The audition tape is your calling card, and it should represent you in the best way possible. It is better not to use a segment than to use a flawed sample of your work. Remember that you may be in competition with broadcasting students and broadcasting professionals from around the country.

The audition tape should be prepared on a new three-inch or five-inch tape. The tape should be of high quality—1.5 mils and a respected brand of tape stock. The materials should be recorded at the standard broadcast speed of $7\frac{1}{2}$ ips. The beginning of the tape should contain leader tape and may include a short section of tone.[1] Exhibit 6.8 demonstrates the proper tape format for all recorded programs including the audition tape.

---

1.   Tone is the audio signal generated from the control board and placed on the magnetic tape. It indicates the 100 percent modulation level and should aid in setting a proper playback level.

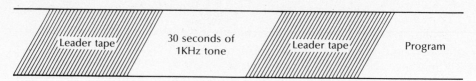

Exhibit 6.8.　Proper Tape Format For Recorded Materials

A résumé can be included with the audition tape and you may conclude your tape with a reminder of your name, your career interests, and information on how you may be contacted. If the tape is unsolicited, it may not be appropriate to request that the tape be returned. The manager did not ask you to send it and may not have time to package it for return to you. If you do want your tape returned, a good idea is to include a return mailer, self addressed and stamped.

It is a good idea to prepare your first audition tape in your junior year of college. This tape can be critiqued by yourself and others and you can work on perfecting it. In your senior year, you should begin your job hunting long before you don the cap and gown for graduation. With proper timing, you can beat the hundreds of other graduates into the job market.

PACKAGING THE FINAL PRODUCTION PRODUCT

Radio production is a job calling for mastery of the technical side of broadcasting as well as the creative side. One exciting aspect of radio production is the ease with which a program can be altered by editing. Cutting segments from a program or cutting segments into a program can change totally the character of the final product. Adding music, sound effects, or mixing sounds can create startling effects.

Every tape should begin with leader tape (about a yard in length). Since tape frequently splinters slightly as it is rewound, the leader tape absorbs the damage and the program material is safe. A 1KH[2] tone recorded at 100 percent modulation placed on the tape provides information about the recording level. The next strip of leader tape indicates that the program begins at the end of the leader tape. The program can be easily cued using this beginning format. A sample drawing as in Exhibit 6.8 can be placed on the outside of the tape box to indicate to the person cueing the tape the exact format that has been used.

The radio producer has a challenging career, as the skills of radio production are mastered. The broadcasting student should experiment with every

---

2.　The 1KH tone is the audio signal generated from the control board and placed on the magnetic tape. The tone is used to set a proper playback level (see note 1).

form of radio production to learn the special production techniques required for each format. Every program should be carefully stored for use in the final student production—the audition tape.

BIBLIOGRAPHY

Alten, Stanley R. *Audio in Media.* Belmont, Calif.: Wadsworth, 1981.

Brady, John. *The Craft of Interviewing.* New York: Random House/Vintage Books, 1977.

Chester, Girard; Garrison, Garnet; and Willis, Edgar. *Television and Radio.* 5th ed. Englewood Cliffs, N.J.: Prentice-Hall, 1978.

Gross, Lynn. *Self Instruction in Radio Production.* Los Alamitos, Calif.: Hwong, 1976.

Hasling, John. *Fundamentals of Radio Broadcasting.* New York: McGraw-Hill, 1980.

Johnson, Joseph S., and Jones, Kenneth. *Modern Radio Station Practices.* 2d ed. Belmont, Calif.: Wadsworth, 1978.

McLeish, Robert. *The Technique of Radio Production.* London: Focal Press, 1978.

Nisbett, Alec. *The Technique of the Sound Studio.* 3d ed. New York: Hastings House, 1972.

Nisbett, Alec. *The Use of Microphones.* New York: Hastings House, 1974.

Oringel, Robert S. *Audio Control Handbook.* 4th ed. New York: Hastings House, 1972.

Orlick, Peter. *Broadcast Copywriting.* Boston: Allyn and Bacon, 1978.

Rosenthal, Murray P. *How to Select, Use, Troubleshoot Cassette and Cartridge Recorders.* Rochelle Park, N.J.: Hayden, 1972.

Seidle, Ronald J. *Air Time.* Boston: Holbrook Press, 1977.

Starlin, Glenn, and Sherriffs, Ronald. *Speech Communication via Radio and Television.* Dubuque, Iowa: Wm. C. Brown, 1971.

3-M Company. *Recording Basics.* St. Paul, Minn.: 3-M Center.

Traylor, Joseph G. *Physics of Stereo and Quad Sound.* Ames, Iowa: Iowa State University Press, 1977.

United States Copyright Office, Library of Congress, Washington, D.C. 20559. *General Guide to the Copyright Act of 1976,* September, 1977.

Walters, Barbara. *How to Talk to Practically Anybody about Practically Anything.* New York: Doubleday, 1970.

Whetmore, Edward Jay. *The Magic Medium: An Introduction to Radio in America.* Belmont, Calif.: Wadsworth, 1981.

QUESTIONS FOR REVIEW AND DISCUSSION

1. What is meant by the "station voice"?
2. What factors contribute to a good radio voice? What factors make a good radio personality?
3. How should the radio announcer prepare for his or her job?
4. What beforehand preparation is needed for a live music program?

5. How does a radio music library function? Why is a music library system absolutely essential?

6. How is a record cued for air?

7. What types of program materials may be used to distinguish one station from another?

8. What must the interviewer be aware of in the interview environment?

9. Briefly outline the steps used in audiotape editing.

10. What should sound effects accomplish in a radio drama?

11. What are the two possible methods for assembling an edited dramatic program? What are the advantages and disadvantages of each method?

12. What is an audition tape?

13. What are the components of a good audition tape?

———————————— *Chapter Seven* ————————————

RADIO STATION OWNERSHIP
AND OPERATIONS

In this country, any individual who is a citizen and who has knowledge of station operations, who meets the Federal Communications Commission's criterion of serving the general public, and who can obtain enough financial backing to maintain the operation, has the potential for owning and operating a radio station.

Anyone interested in a career in radio should understand something about radio's daily operations. This chapter examines the radio station and explores such topics as: what are the requirements for radio station ownership? How are radio station allocations made? How are radio stations organized? What do the various positions at a radio station involve — management, sales, news directing, traffic, and others? How does a station select a format? What factors define a successful radio station?

As we have noted in earlier chapters, radio has changed over the years from a national medium to a local medium. One radio practitioner has observed that individuals "rarely remember the names of radio shows nowadays, but most people can recognize the sound of a favorite station even when dialing for it blindfolded."[1] What is a station "sound" and how is a sound developed and maintained? To answer these questions, let's start at the beginning and explore the avenues to radio station ownership.

1. J. Leonard Reinsch and Elmo Israel Ellis, *Radio Station Management,* 2d ed. (New York: Harper, 1960), p. 1.

137

PREREQUISITES FOR RADIO STATION OWNERSHIP

The License Requirement

Understanding how an individual or group of individuals obtain a broadcast station is essential to understanding the communication system in this country. In some countries, broadcasting is the right of the privileged. Anyone interested in broadcasting must be wealthy and friendly to those who rule the country. This is not the case in the United States. In some countries, the citizens do not understand how an individual acquires the right to broadcast because such information is classified. In this country, any individual who is a citizen of the United States and who has knowledge of station operations, who meets the Federal Communications Commission's (FCC) criterion of serving the general public, and who can obtain enough financial backing to maintain the operation, has the potential for owning and operating a radio station.

To operate a radio or television station in the United States requires a license from the federal government through the FCC. There is limited frequency space, which means that not everyone can operate a station. If broadcast operations were not licensed, it would be impossible to receive a clear radio signal because the various unregulated signals would overlap and interfere with good reception. The Communication Act of 1934 decrees that every radio station must be licensed before broadcasting. The license specifies the frequency on which the station can operate, hours of the day that broadcasting is permitted, the amount of power of its signal, and other conditions of operation.

The act also decrees that, since the frequency spectrum is limited, broadcasting is a *privilege*. Therefore, anyone receiving a broadcast license must agree to operate in "the public interest, convenience, and necessity." This phrase is the landmark of American broadcasting and characterizes commercial station operations in this country.

How does a would-be radio station owner acquire a radio license and prove a willingness to operate in the public interest, convenience, and necessity? There are two general types of radio stations—AM and FM. Let's first consider the route to acquiring an AM station.

AM Station Ownership

The FCC regulates over 4,500 AM stations operating on 107 carrier frequencies, which begin at 540 kilocycles and are successively placed with a separation of 10 kilocycles up to 1600 kilocycles on the dial. The allocations for AM frequencies are authorized in three major categories—clear channels, regional channels, and local channels.

Anyone interested in owning a new AM station must conduct a thorough study of the coverage area (both daytime and nighttime) of all other stations that may interfere with the proposed station or be subject to interference from the proposed station. Other AM applications pending at the FCC must also be checked for possible interference problems. The FCC will check only the conclusions once they are incorporated into a station license request.

The frequency search has become so complicated (by the large number of existing AM stations and by AM's characteristic of being carried at great distances by skywaves in the evening hours) that many would-be station owners have employed broadcast consultants to do the job for them. These consultants use the computer to insure that they have not overlooked any station that may have an interference problem, and their services are usually costly. The problems associated with starting a new AM operation are frequently prohibitive for the individual with limited resources, but purchasing an existing AM station may be less taxing on the physical and financial resources.

FM Station Ownership

Since FM stations are not subject to having their signal carry over great distances by skywaves or groundwaves, all FM stations are licensed for unlimited time of daily transmission. The technical portion of the initial application to the FCC is much less demanding than the application for an AM station. The FCC maintains a list of available FM frequencies (FM Table of Assignments), and a would-be owner can check the available frequencies in the desired area of operation. No major frequency search is required, making the preliminary investigation for an FM license less costly than for an AM license.

The FM spectrum is divided into zones, with Zone I being the most populous parts of the country—portions of eighteen northeastern states and the District of Columbia; Zone IA is Southern California, Puerto Rico, and the Virgin Islands while Zone II is the rest of the United States, Alaska, and Hawaii.

The FM allocations are further divided into classes that recognize the two factors affecting FM station coverage—the height of the transmitting antenna and the "effective radiating power" (ERP). Since the FM band is in the very high frequency range, the FM signal is transmitted primarily by directwaves unaided by groundwaves or skywaves. Directwave transmission is also called *line of sight* transmission because the range is theoretically limited to the terrain that would be visible from the top of the antenna tower. The height of the terrain surrounding the FM antenna is important since the higher the FM antenna above the surrounding terrain, the greater is the station coverage area.

The "effective radiated power" (ERP) is the power gained by reflectors that redirect waves that would have gone toward the sky back down toward the horizon. Since FM radio waves are not subject to skywave transmissions, these emissions would have been lost if not reflected back toward the horizon. These reflected waves increase the power of the FM signal, making the ERP an important factor in station coverage.

The FM stations are divided into classes that are defined by ERP and antenna height. A Class A station is intended to render service to a relatively small community with an ERP of 100–3000 watts and an antenna height of not more than 300 feet above the average terrain. A Class B station is designed to serve middle-sized cities with an ERP of between 5000 and 50,000 watts and a maximum antenna height of 500 feet. A Class C station is designed to serve a large population area with an ERP of 25,000 to 100,000 watts and a maximum antenna height of 2,000 feet above the average terrain.

FM station operations are particularly attractive because the frequency width allocated for each FM station permits additional signals to be transmitted on the same channel. *Multiplexing,* which means transmitting more than one signal in the same channel allocation, has made FM stereo broadcasts possible and is a means through which FM stations can increase their revenues. Through multiplexing, FM stations can provide "muzak" services to subscribers in the business community or lease time for weather and local information channels. These extra benefits, which can be developed by an FM station owner, make an FM station particularly appealing for the investor who must maximize the financial potential of the station.

Ownership Essentials

An application for permission to construct a new broadcast facility or transfer ownership of an already authorized facility must be requested from the FCC as the first step in broadcast station ownership. On this form, the applicant is asked to identify all of the proposed owners of the station, percentages of ownership for each partner, and all pertinent information regarding moral turpitude, history of other business dealings, and any interest in or connection with any other broadcast operation of all the proposed partners. Information regarding the business, personal, and professional life of the proposed applicant(s) is considered vital information, as the FCC attempts to license station operators who will serve the "public good."

The individual(s) proposing the station must also include information regarding the type of programming that will be provided to the local community. The FCC does not want to control broadcast programming in the United States, but the commission does want to guarantee that the station will serve the local community.

In recent years, the FCC has decided that marketplace forces should dictate station formats. Local citizens will listen to the station that provides programming that meets the needs and interests of the listeners. The marketplace theory is new at the FCC. The commission used to be very much concerned with station programming and the percentage of total programming devoted to news and public affairs. Now, the commissioners believe that marketplace forces—the listening public and competition from other stations and other media—will dictate quality programming.

The license application form also has considerable space for information regarding the station equipment and engineering aspects of the proposed station. The applicant must be very careful to complete the engineering information correctly, or the application may be disqualified for incomplete or inaccurate information.

When two or more parties are competing for the same facility, the applicants are equally qualified as station owners, and this is a first station for the applicants, the FCC may, in the future, use a lottery system to select the new owner.[2] In more complicated cases, a hearing is usually set in Washington, D.C., where an FCC examiner hears arguments from all parties and considers the facts. The hearing examiner reports the findings back to the commission, which deliberates and makes a decision based on the facts. The FCC attempts to award the license to the applicant who can best serve the local community. The parties to the hearing have the opportunity to contest the decision by carrying the dispute to the federal courts.

When the application has finally been processed and approved, the applicant will be issued a construction permit (CP). The transmitting facilities must be built *exactly* as approved in the application. The FCC field engineer's office inspects the construction of the station's transmitter. The station cannot go on the air until the field engineer's office has approved the transmitting facilities.

The new station must select call letters that are not the same as the call letters for any other station in the United States. By an international agreement, letters K and W were assigned to the United States to be used as the first letter of the calls of each broadcasting station. It was further decided that stations with K lead-off would be located west of the Mississippi and that stations with W as the first letter would be located east of the Mississippi. Before the geographic boundaries were drawn, some stations already had call letters that were in conflict with this ruling. These stations were allowed to keep their calls, giving us such stations as KDKA in Pittsburgh and WHO in Des Moines. Three-letter identifications were originally assigned to stations, but three-letter identifications exhausted the alphabet so a fourth letter was added.

Once the applicant has selected the call letters and constructed the transmitting facilities and studios, the applicant must continue to report to the FCC. Reports that each station must file include: annual financial report, any

2. The FCC is considering the use of a lottery system, but some problems must be solved before the lottery can go into effect. Special consideration should be given to station ownership requests from minorities. How will a lottery provide this "special consideration"? This question and others must be answered before the FCC institutes the lottery system. The trade publications will alert those interested in broadcast ownership of the continued debate over the lottery issue.

changes in ownership, any modifications that the applicant would like to make in the license agreement, requests for license renewals, and any other reports the FCC may solicit.

Purchasing a Radio Station

A good avenue to radio station ownership may be to purchase an existing station. There is no formula for determining the value of one, but there are some factors that should be considered before agreeing upon a price.

> Power and frequency
> Economic condition of the station
> Reputation of the station
> Calibre of the staff
> Condition of the equipment, studios, and transmitter building
> Competition in the local market
> Outside competition
> Network affiliation
> Present and long-range prospects for the market
> Possible new competition
> The effects of technical advancements on the present facility[3]

Once these considerations have been made, an equitable price for the station can be determined. The difficulty in assessing price is that no hard and fast dollar value can be placed on some of the considerations. The reputation of the station must be considered, but how much, in dollar terms, is a good reputation worth? How much should be deducted if the station has some difficulties with its reputation in the local community? Such considerations make assessing a dollar value very difficult.

In purchasing a radio station, the services of a good attorney are absolutely essential. Law firms that specialize in legal services to broadcasters are located in Washington, D.C. It is not uncommon for a broadcasting station to utilize the services of both a local attorney and an attorney located in the nation's capital. Purchasing a radio station requires careful consideration and utilizing the service of someone with experience in this form of transaction is essential.

One final point needs to be made about radio station ownership: the price of station ownership continues to increase. In days gone by radio stations could be purchased for thousands of dollars; now the price tag frequently reads in the millions. Some recent examples include: KMJK–FM, Lake Oswego, Oregon sold for $2,200,000; KNDE–FM, Tucson, Arizona sold for $2,600,000; KOOK–AM–

---

3.   See Reinsch and Ellis, p. 16.

FM, Billings, Montana sold for $2,005,500; KBIL–FM, San Angelo, Texas sold for $1,600,000; KDUV–FM, Brownsville, Texas sold for $1,019,000; and KQPD–FM, Ogden, Utah sold for $760,000.[4] *Broadcasting* magazine is an excellent source of information on radio station sales.

INSIDE THE STATION

The Staff

Station ownership can take many forms. The owner can be a lone individual or the owners may be two or three partners, a group of individuals, or a large, impersonal corporation. Regardless of who owns the station, the FCC holds the owner(s) responsible for station operations, since the license is issued to the owner(s).

The station owner(s) insures good broadcasting by employing a highly qualified and dedicated staff. Most stations have a station manager, station announcers, someone who heads a "traffic" department, at least one engineer, and sales personnel. The number of employees varies from station to station and depends upon the market area of the station. A large station may employ thirty or more employees while a small market station may employ just two or three. The various jobs that may be included in the station staff are listed below.

The Management Staff

The station manager, who is also frequently the station owner, is the individual whom the public identifies as primarily responsible for station operations, including programming. In radio, the station manager most frequently comes to the job by way of the sales department. A thorough understanding of sales is often considered a pass card to commercial broadcast management.

The station manager is responsible for all the station's operations. The manager is expected to be a leader who inspires the sales department to bring in new accounts, challenges the production staff to be creative and demanding in their programming, and encourages the engineers to keep abreast of the latest radio technology and to maintain excellence in the station's equipment. In addition, the station manager must maintain contact with the local community. In short, the responsibility for good station operations rests with the manager.

The jobs of the manager have been categorized into the following divisions:

1. Setting objectives for the station
2. Determining the tasks needed to accomplish these objectives

4. *Broadcasting,* October 17, 1983, pp. 71–72.

3. Assigning responsibilities and authority for meeting these objectives
4. Supervising employees in the process of meeting objectives.[5]

The production manager is responsible for the station's programming. The individuals who are accountable to the production manager frequently include: program researchers, staff announcers, editors, copywriters, and members of the traffic department.

The news director is responsible for all news operations at a station. This individual recruits and hires reporters, writers, and sometimes production personnel to assist in news programming. The news director establishes regular news beats for the staff and establishes a news format and a news style.

The sales manager is responsible for the activities of the sales and promotional personnel. The sales manager must maintain up-to-date facts and figures about the listening audience. This information will enable the sales department to establish advertising rates for the station. With this information the sales personnel go into the business community and attempt to sell the station's time. Chapter Eight discusses the specific jobs of the sales and promotion department.

The Office Staff

The administrative assistant helps the station manager by keeping the books and by assisting in the gathering and organizing of information that must be sent to the FCC. The administrative assistant may also may keep the manager's calendar.

The office manager is responsible for supervising the office staff— secretaries, receptionists, clerks, and sometimes janitorial help. The office personnel are generally charged with typing, answering the telephones, collecting the mail, and performing the other duties that are essential to an efficient and pleasant working environment.

Engineering, Traffic, and Subdepartments

The chief engineer is responsible for installing and maintaining all broadcast equipment at the station and at the transmitter site. At a large station, the chief engineer will have three to ten other engineers in the engineering department; at a small station, the chief engineer may be the only engineer. A very small station may employ just a part-time engineer. The chief engineer generally reports directly to the station manager.

The traffic department is responsible for maintaining the station's daily programming log. This essential job involves maintaining a complete record of the

5. Reinsch and Ellis, p. 16.

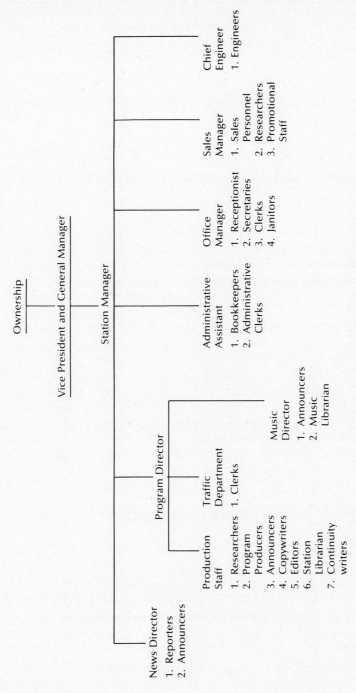

Ownership

Vice President and General Manager

Station Manager

Program Director

News Director
1. Reporters
2. Announcers

Production
Staff
1. Researchers
2. Program
   Producers
3. Announcers
4. Copywriters
5. Editors
6. Station
   Librarian
7. Continuity
   writers

Traffic
Department
1. Clerks

Music
Director
1. Announcers
2. Music
   Librarian

Administrative
Assistant
1. Bookkeepers
2. Administrative
   Clerks

Office
Manager
1. Receptionist
2. Secretaries
3. Clerks
4. Janitors

Sales
Manager
1. Sales
   Personnel
2. Researchers
3. Promotional
   Staff

Chief
Engineer
1. Engineers

Exhibit 7.1.   An Organizational Chart for a Large Market Radio Station

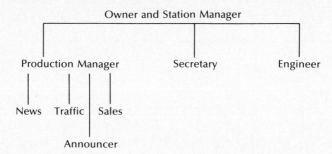

Exhibit 7.2.   An Organizational Chart for a Small Market Radio Station

commercials and programming and all announcements for each day of broadcasting. The log serves as a set of instructions for the station personnel in getting material on the air. The log also provides a record of broadcasting performance for accounting, FCC reports, and other purposes.

The music department may form a subdepartment under the production department. The music director is responsible for the musical selections that are aired by the station. This individual, in consultation with station management, selects a "sound" for the station. The music director supervises the work of the music librarians and the work of the music announcers.

At small stations a few staff members may serve many functions. The employee of one small market station reported that his station was a one-horse operation and that he was the horse. This employee served as program manager, traffic department, announcer, news director, and station engineer. Organization of any station will depend on the needs of the particular station. Exhibit 7.1 shows a typical organizational chart for a large station and Exhibit 7.2 shows such a chart for a small station.

THE STATION FORMAT

Creating a Sound

Each radio station must decide on the "sound" for that particular station. In order to attract an audience, the station must have a unique format with which the listener can identify. Station formats have included:

All news
Country music
Christian broadcasting
Country and western music
Middle of the road (MOR)

Beautiful music
Rock
Black (soul format)
Classical music
Oldies
Educational
All talk
Contemporary

Selecting a station format is one of the most crucial decisions that a station must make.

Decision Factors

Management must consider a variety of factors before the final decision is reached in the format selection process. The local community must be surveyed to determine what the potential audience wants to hear. The survey should include facts on the educational levels of the potential audience, economic background, social concerns, and information on any other factors that would be significant in directing programming and advertising materials to the audience.

Competition from other stations must be evaluated. A small community usually cannot sustain more than one type of station—one country or one rock, for example. A well-established station can create difficulties for any new station that moves into the area and goes after the same audience the older station has built and carefully maintained over the years. Also, the FCC likes diversity in programming and may be reluctant to license a station only to give the local community more of what it already has.

The size of the community is also a factor in selecting a format. Research indicates that an all-news format has its best chance of survival in a large market. A classical music format may find its place in a market with a large number of individuals in mid-to-upper income groups with college degrees. Most stations in small communities develop an MOR format that appeals to the widest possible audience.

The key factor in format selection is: can revenues be generated for the station if it adopts a particular format? If a station wishes to adopt a Christian music format, for example, can revenues be obtained to keep the station on the air? If the station is noncommercial, can grants and donations be obtained to support the format?

The management must also consider its own interest in the format and the abilities of the station personnel to successfully coordinate the format. A radio staff that dislikes the station sound is not likely to generate quality programming.

Also, in selecting a format, the management must consider how public service programming will fit in. The FCC may not look favorably on a station that moves into a market and fails to provide public service programming.

And, finally, after the format has been selected, the audience must be continually surveyed to detect changes in their interests and tastes that would affect the format. A station sound that is not updated may gradually lose its audience to other stations more attuned to community changes.

Syndicated Formats

Today many radio stations hire a program consulting firm to select a format for the station through survey research of the community and through careful analysis of national and regional trends in musical tastes and entertainment preferences. These research/consulting firms usually offer the station a complete format, which includes musical selections, promotional concepts, station identification jingles, and research information on the local community. Estimates indicate that approximately one-half of the commercial radio stations are subscribing to some form of consulting firm for syndicated programming.

The syndication companies charge for the package according to the options that the station selects. Most of the syndication companies offer a beautiful music format, rock/contemporary formats, and an MOR format, and many of the companies have a country music format to sell. Soft rock and album-oriented rock are also common offerings. The concept of the syndication/consulting company that provides the station sound is a recent development. More than half of the radio syndication companies were formed since 1973. Millions of dollars were spent on radio syndication/consulting last year. The syndication companies have been so profitable that now numerous companies offer syndicated programming. Exhibit 7.3 lists some of the radio syndication companies located in the United States. The oldest and possibly best known of these are Drake-Chenault and Bonneville.

Most of the stations using a syndicated format also use automated equipment. Some syndication companies send the station musical selections on ten-

American Music Research	Drake-Chenault Enterprises
Bonneville Broadcasting System	Musicworks
CaVox Productions	Peters Productions
Century 21 Programming	Radio Arts
Churchill Productions	The Creative Factor
Concept Production	TM Programming

Exhibit 7.3.   Some Radio Program Syndication Companies

Source: The Radio Programs Source Book, 2d ed. A publication of Broadcast Information Bureau, 1983.

Exhibit 7.4.   An Automated Radio System

At an automated station, the programming is computer controlled. The computer is programmed to play tapes and carts at precise times throughout the broadcast day. Many computer-controlled (automated) stations use the services of a radio programming syndication company. In the control room of an automated station, the cart and tape machines start and stop without direct human intervention. The control room is alive with automated activity, but human intervention entered the scene only when the computer was programmed for the day's broadcast schedule. Photos courtesy of KEZT, Ames, Iowa.

inch tape reels and station IDs and jingles on carts. The station computer is programmed to play the various tapes and carts at precise times. Exhibit 7.4 shows an automated radio system. Some syndicated companies and some networks are sending their programs by satellite.[6] The syndication companies and networks are supplying or leasing the station earth receivers to pick up programming.

The syndication companies insist that they work with the personnel from the local station to give the station a unique sound. As a result, the syndication companies claim that no two stations should sound alike.

The list in Exhibit 7.3 indicates just some of the radio program syndication companies that offer services to radio stations. Some offer complete programming services to the stations—audience research, format selection, complete formats on tape, station identifications and jingles, and promotions. Other syndication companies offer only specialized programs—news, public service, or special music programs.

Music Licensing

Some 80 percent of all radio station programming in this country consists of music. One of the major operating costs associated with a music format is the music license fee. Every station must have clearance from a music licensing organization and pay a fee to play each record that is aired. Music licensing is controlled primarily in the United States by three organizations—The American Society of Composers, Authors and Publishers (ASCAP); Broadcast Music Incorporated (BMI), and The Society of European Stage Authors and Composers (SESAC).

ASCAP, which was formed in 1914, is the largest and the oldest. ASCAP now has the rights to over a million titles, and contracting for music rights with radio stations accounts for about four-fifths of its annual receipts. BMI was formed in 1940 as a music licensing agency for new composers who could not afford to affiliate with ASCAP. BMI now has rights to over a million titles and is second only to ASCAP. SESAC was formed originally to represent European artists but is now based in the United States. SESAC was formed in 1931 and is the smallest of the three music licensing organizations. All of these organizations have had major skirmishes with the broadcasters who frequently resent having to pay for the privilege of airing a record when the very process of broadcasting the record benefits both the artist and the recording companies. These skirmishes will probably continue, but at least at the present, the music licensing organizations are well entrenched and the radio station must acquire the right to broadcast music.

---

6.   ABC, CBS, NBC, Mutual, and other networks are currently using satellite delivery systems. See
      Broadcasting, January 2, 1984, p. 60.

The station pays a general fee to the music license company in order to play the titles held by the company. If the station does not subscribe to a syndication company for its format, the station must negotiate with the music licensing companies itself. Some of the syndication companies have already negotiated the music contract on the formats they sell. The fees to the music licensing companies are included in the total costs charged to the station by the syndication company.

Some music, such as folk music, is in the public domain. This means the music can be aired without paying a fee to any company. Most stations play popular music, however, which is licensed by ASCAP or BMI.

Music Selection within the Station

In stations where music is not prepackaged by a syndication company, musical selections may be made in several ways. In some stations the music is selected by the station manager; in others, it is the job of the program director; still other stations leave the selection to the DJs; some hire musical directors who select the music to be aired; and some rely on national lists or local lists of best selling records.

The practice of receiving benefits for selecting music to be broadcast caused the payola scandals of the late 1950s. Evidence showed that, in some stations, persons were given money or merchandise to play certain music. Music aired on large market stations has a high probability of earning large revenues from record sales; thus the recording companies are always eager to have their new albums aired. The payola scandals surfaced again in the 1970s when DJs in Washington, D.C., were accused of programming for pay. The selection of music for a station is serious business and directly affects the station's relationship with the audience, the recording companies, and the FCC.

Other Ingredients of the Format

Radio drama has become an important feature at some stations. The broadcasting trade publications report that many stations have an avid audience for radio drama. Reruns of early radio dramas including "Sam Spade," "Ellery Queen," "Sherlock Holmes," and others are now available through syndicators and are a welcome addition to many formats.

Other ingredients of the station sound include weather reports, talk programs, news, public service programs, and feature programs. The talk format is popular in some markets and usually consists of programs that invite the listeners to call in and offer opinions or information. Some programs on talk radio give the listeners a chance to comment on issues in the news, others provide a chance for sports enthusiasts to sound off, and still others have included listener exchange of recipes, community events, or travel information. Some programs provide listeners an opportunity to interact with guest "experts." Some talk stations also include features provided by syndication companies or features produced by the programming staff.

Another segment of any basic format is news. The trade publications indicate that the outlook is good for a variety of syndicated news sources featuring people-oriented programs and news background materials. These services offer the listener more than just straight news.

Whatever the format, the station can maintain its own unique sound if the station manager and the station's production team utilize the opportunities to make the station unique. Remember that the format is composed of the entire station sound—music, news, features, talk programs, announcements, advertisements, and the personality conveyed by the station announcers. The individual components of the total format package distinguish one station from another. By mixing various ingredients into a total package, unique station sounds can be created.

THE SUCCESSFUL RADIO STATION: KEYS TO SUCCESS

There is no one kind of program format that can be classified as a surefire ticket to success. The key to success is in knowing the audience and working to please the audience.

The general belief in radio today is that almost any format can lead to success if the station properly packages and markets it. The great diversity of "winning" formats supports this belief. Several years ago the top format in the New York City area was urban contemporary, in Los Angeles it was talk, in Chicago it was MOR, while in Kansas City it was country.

Many radio practitioners believe that in a small market, a radio station must be diverse, offering a variety of music, news, and features. A diverse station sound provides a great opportunity for creative management and for a creative production staff. The formats that lead to success in radio, whether in small markets or large, are based on hard work, close community ties, creativity, competent management, and a well-trained and dedicated staff.

MANAGEMENT CAREERS: OPTIONS

Most broadcasting students think of broadcast production as a career choice rather than station management or ownership, which often appear to be "pie in the sky." This attitude is unfortunate, since station management and ownership are within the reach of almost any U.S. citizen who properly plans for the opportunity.

The difficulty of interesting students in station management is illustrated in the following story. A few years ago Lorenzo W. Milam published a book titled

Sex and Broadcasting. The book had a lavender cover and unusual drawings, and it boasted "straightforward, unexpurgated language." The subtitle of the book was "A Handbook for Starting a Radio Station for the Community." Milam used his unusual title to attract readers because a handbook on starting a radio station is considered dry reading by most broadcasting students. The marketing gimmickry used by Milam illustrates the difficulties of interesting students in station management and ownership.

Broadcast station management can be as creative in its approaches as can be broadcast production. The challenge for the future is to attract to station management those individuals who have an understanding of the broadcasting industry and a desire to offer the public the very best that radio can provide. Many station managers do not utilize the opportunities to develop unique formats to serve the community. One radio critic has complained that most radio stations sound alike regardless of the community. Too few broadcasters are willing to experiment and insist on relying on programming that has been tested at other stations. Station management offers exciting career possibilities for the student who looks ahead and prepares for the challenge. Radio stations do not have to sound alike, but creative management must be introduced into the broadcasting industry to maintain audience support for radio.

BIBLIOGRAPHY

Barcus, Ray. *You as a Broadcast Employee.* Dallas: Elkins Institute, 1970.

Broadcasting magazine, 1983.

Coddington, Robert H. *Modern Radio Broadcasting: Management and Operation in Small-to-Medium Markets.* Blue Ridge Summit, Pa.: TAB Books, 1969.

Coleman, Howard W. *Case Studies in Broadcast Management.* New York: Hastings House, 1975.

Ellis, Elmo. *Opportunities in Broadcasting.* Skokie, Ill.: National Textbook, 1977.

Finnegan, Patrick. *Broadcasting Engineering and Maintenance Handbook.* Blue Ridge Summit, Pa.: TAB Books, 1976.

Hasling, John. *Fundamentals of Radio Broadcasting.* New York: McGraw-Hill, 1980.

Hoffer, Jay. *Managing Today's Radio Station.* Blue Ridge Summit, Pa.: TAB Books, 1968.

Johnson, Joseph S., and Jones, Kenneth K. *Modern Radio Station Practices.* 2d ed. Belmont, Calif.: Wadsworth, 1978.

Milam, Lorenzo W. *Sex and Broadcasting: A Handbook on Starting a Radio Station for a Community.* Saratoga, Calif.: Dildo Press, 1975.

Quaal, Ward, and Brown, James A. *Broadcast Management,* 2d ed. New York: Hastings House, 1976.

The Radio Programs Source Book, 2d ed. Broadcast Information Bureau, 1983.

Reinsch, Leonard J., and Ellis, Elmo Israel. *Radio Station Management.* New York: Harper, 1960.

Whetmore, Edward Jay. *The Magic Medium: An Introduction to Radio in America.* Belmont, Calif.: Wadsworth, 1981.

QUESTIONS FOR REVIEW AND DISCUSSION

1. What are the basic prerequisites for radio station ownership in the United States?
2. What is the landmark of American broadcasting that characterizes commercial station operations in this country as unique from any other country?
3. What are the three general classifications for AM radio stations?
4. Why is a frequency search required for a new AM station? What factors contribute to the complexity of the AM frequency search?
5. What factors make an application for a new FM station easier to file than for an AM station?
6. What two factors affect FM station coverage?
7. How is multiplexing used to make an FM operation more profitable?
8. What type of information is required on a radio station application?
9. What restrictions govern the selection of station call letters?
10. What factors should be considered before agreeing upon a purchase price for an existing station?
11. What are the responsibilities of the station manager?
12. What are the responsibilities of the traffic department?
13. What factors should be considered in determining the size of and the organization of the station staff?
14. What factors should be considered in selecting a station format?
15. What are some of the leading radio syndication/consulting companies?
16. What are the three largest music licensing companies in the United States?
17. What is a station sound?
18. What types of radio formats have proven successful in the United States?
19. What factors lead to a successful radio station operation?

————————————— *Chapter Eight* —————————————

RADIO SALES
AND PROMOTION

*The major concern of any radio station is generating enough
money to remain on the air. Generating dollars for the station
is the primary responsibility of the sales and promotion
department.*

The major priority for any radio station is earning enough money to remain on
the air. The sales and promotion department is generally concerned with
building a large and devoted audience for the station and selling air time to
advertisers. This chapter explores the jobs of the sales and promotion department and examines the avenues to building a loyal audience and generating
revenues for the station.

THE SALES DEPARTMENT

Advertising and Promotion

At most stations, sales is considered one of the most vital jobs at the station.
The sales and promotion department is charged with the responsibility of
recruiting advertisers for the station and with promoting the station to listeners.
To sell the station's air time, the salesperson approaches the potential advertiser with very specific information about the station.

1. The number of listeners in the station's total audience.
2. The median age of the station's listeners.
3. The number of listeners who tune in the station at a particular time of day.
4. The type of programs aired throughout the broadcast day.

AFFLUENCE:

Ames ranks

138th in Consumer Spendable Income per
household and

Ames ranks

186th in Total Retail Sales per household
yet in population *318th*.

*Source Dec. 1978 Edition SRDS

KASI

The only commercial AM station in Story
County.
KASI targets the 25+ age group with
genuine, adult M.O.R. music.
KASI concentrates heavily on local, Ames
news, sports and information.
KASI news sources include: United Press
International, UPI Audio, the Iowa Radio
Network, Freese-Notis Weather Service
plus its own full-time, local news staff.

Ames

Home of Iowa State University and the
ISU Center, the cultural hub of Central
Iowa.

KASI

1000 Watts at 1430 KHZ

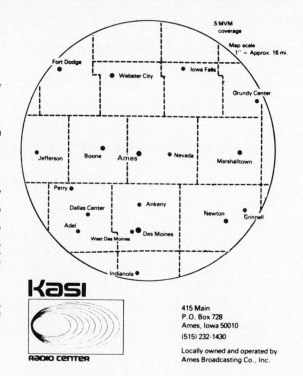

KASI

RADIO CENTER

415 Main
P.O. Box 728
Ames, Iowa 50010
(515) 232-1430

Locally owned and operated by
Ames Broadcasting Co., Inc.

Exhibit 8.1.
These examples are promotion materials that stations KASI (*above*) and KCCQ (*opposite*), Ames, Iowa, prepared to attract advertisers. The sales staff tries to convince the advertiser that benefits will be gained by advertising on the station. (Photos by permission of KASI/KCCQ.)

5. Unique information about the station's listening audience—family income, products used by the listeners, lifestyle of the listeners, and other types of specific data.

With this type of information, the salesperson tries to convince the potential advertiser that investing in radio advertising will be profitable. The advertiser's cost is expressed in CPM, cost per thousand. *Cost per thousand* is the amount of money required to reach a thousand persons with an advertising message. (The M in CPM stands for mil, the Latin term for thousand.)

The concept of CPM becomes a little complicated when the advertiser aims a message at a very specific target audience. Let's say a manufacturer of support pantyhose wants to buy radio air time on a Chicago station at 11:30 AM. The advertiser decides to invest $1,000 in the spot. At 11:30 AM, the radio station has 10,000 total individuals in the audience, making the CPM $100. But, the advertiser is interested in reaching only the women who, let's say, compose one-half of the listenership at that specific time. The CPM for the advertiser then becomes $200. But even more specifically, the support hose manufacturer

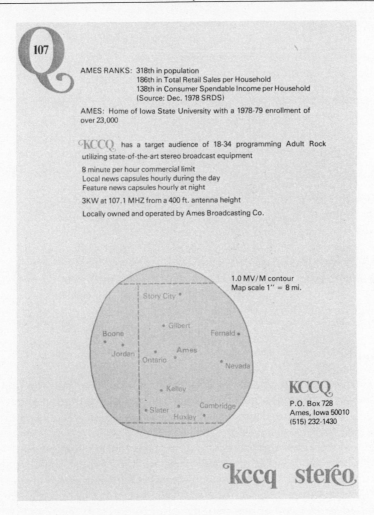

AMES RANKS: 318th in population
186th in Total Retail Sales per Household
138th in Consumer Spendable Income per Household
(Source: Dec. 1978 SRDS)

AMES: Home of Iowa State University with a 1978-79 enrollment of over 23,000

KCCQ has a target audience of 18-34 programming Adult Rock utilizing state-of-the-art stereo broadcast equipment

8 minute per hour commercial limit
Local news capsules hourly during the day
Feature news capsules hourly at night

3KW at 107.1 MHZ from a 400 ft. antenna height

Locally owned and operated by Ames Broadcasting Co.

1.0 MV/M contour
Map scale 1″ = 8 mi.

KCCQ
P.O. Box 728
Ames, Iowa 50010
(515) 232-1430

kccq stereo

really wants to reach women between the ages of forty and fifty. There are 2,500 of such women in the listening audience, making the CPM $400. The CPM is derived by dividing the costs of getting the advertisement on the air by the number (in thousands) of target individuals in the listening audience.[1]

The job for the radio salesperson, then, is to demonstrate that the station can reach the target audience at a reasonable CPM. The salesperson never approaches a client without having specific information about the listening audience so as to entice the advertiser to buy station time. Exhibit 8.1 shows promotion materials that one station prepared to attract advertisers.

1. The above example is used just to illustrate how to calculate CPM. The example is not realistic in that one radio spot would not cost $1,000. The advertiser would purchase a series of spots to run over several months. The CPM would be much less costly because the advertiser would reach some new listeners each time the spot ran.

Advertising Options

When purchasing air time, the advertiser may select one of several options. The advertiser may decide to become a *regular sponsor* of a particular program. This means that the advertiser's spots run during a particular program and the advertiser becomes identified with the program itself. A station may carry a travel program, for example, during which time listeners call in to ask questions of travel experts about vacation spots or weather conditions in a location where they plan to travel. A travel agency may decide to become a regular sponsor for the program, thus providing the agency with an excellent target audience week after week.

Another type of buying is called *scatter spots*. The advertiser may select times throughout the program week for spots of varying lengths. This allows the advertiser to examine the station's audience demographics and to select various times when large numbers of potential product purchasers are in the audience. Most air time purchases are scatter spots.

The advertiser may decide to buy *station break spots* that run between programs. The advertiser could still select the time of day for the spots, but they would not be identified with any particular program.

Rate Cards

Rate cards are used by stations to supply advertisers with information about the station's advertising rates. The rate card provides information about station coverage, time classes, audience size, and audience demographics for specific times during the broadcast day.

Station coverage is the total geographic area that the station reaches. This information is particularly important for small market radio, which may reach several towns though the station is primarily identified with only one small town. The total coverage area for a small market radio station may impress a potential advertiser and entice the advertiser to purchase air time.

Audiences for Radio

Most radio stations have what is called *drive time*—the time between 7:00 and 9:00 AM when listeners are driving to work, and between 4:30 and 5:30 PM when listeners are driving home. During drive time, the station's audience is usually at its peak. During drive time, radio advertising rates are generally higher than for any other time of day. Drive time may vary from city to city; in Washington, D.C., for example, it is from 7:00–10:00 AM and from 3:00–6:00 PM. The drive time hours are extended because federal employees have staggered work schedules.

A station generally divides the day into classes that are representative of audience size and audience composition. An example of a station's time divisions is: 6:00–10:00 AM, 10:00 AM–3:00 PM, 3:00–7:00 PM, 7:00 PM–Midnight, Midnight–

7:00 AM. These classes may represent very different audiences. The 6:00–10:00 AM segment may be the station's largest audience, composed predominantly of working men and women. The 10:00 AM–3:00 PM division may be predominantly housewives and retirees who are in the home during the daytime hours. The 3:00–7:00 PM division may be predominantly teenagers who are out of school in the afternoons and use radio while doing homework or talking on the telephone. If the station covers local sports in the evening hours, the 7:00–Midnight slot may be primarily sports fans. The time classes with accompanying advertising rates are determined by the number of listeners and the specific type of listeners during that time of day.

Over the years, the characteristics of the radio audience have changed dramatically. Peak listening hours for radio used to be evening hours from 7:00–11:00 PM, which are now television's prime time hours. Now, daytime hours are frequently reserved for radio listening when a station can be tuned in at the office or in the home while the listener engages in other activities. For most radio stations, the most expensive air rates are for daytime hours, but some stations are using only one rate for the entire broadcast day. Most stations, however, continue to divide the day into times characterized by audience size and audience demographics and to charge rates based on these divisions.

Audience Demographics

Audience demographics are specific pieces of information about the listeners who tune in at a given time. The information that is most frequently included in radio listenership demographics are: age of listeners, sex, marital status, income, education, work habits, purchasing habits, and lifestyle information. This type of information helps an advertiser decide what time of day to run an ad.

A manufacturer of blue jeans may want to purchase time when large numbers of middle-to-upper-income teenagers are in the listening audience. At station WXXX, that time may be from 3:00–7:00 PM. Station WXXX may have a predominantly female, middle-aged audience from 10:00 A.M.–3:00 PM that would not be interested in the manufacturer's new line of hip-hugging jeans designed especially for extra-slim figures. During other times of the day, the station's audience may be equally inappropriate for this advertiser. Demographic information about the station's various audiences would be included on the rate card to indicate to this manufacturer exactly when a specific target audience can be reached.

Quantity Discounts

In addition to varying rates for different times of the day and for different types of audiences, the station may also offer special rates. One type of special rate is the *quantity discount.* Many stations offer discounts if the advertiser purchases a large number of spots. The rate structure for quantity purchases is completely a station management decision. The station may decide to offer a percentage cut for quantity purchases: a 10 percent discount to any advertiser who purchases a minimum of 100 spots, with a 20 percent discount for customers who purchase 500 spots, and a 30 percent discount for 1,000 spots. The cutoff points and the alteration of the rate structure are determined by each station. This type of information is frequently indicated on the rate card to encourage the advertiser to purchase advertising time in quantities.

Actually, the quantity purchase benefits both the station and the advertiser. The advertiser is most likely to reach the largest target audience if the ads run over an extended period of time. Thus the quantity purchase aids the advertiser. This arrangement also helps the station by allowing the station to plan its cash flow policies and air time schedule.

Local vs. National Advertising

Another rate differential is usually applied to the local vs. the national advertiser. Traditionally, the national advertiser must pay more for a station's air time than a local advertiser. This is because the national spot costs the station more in overhead than does the local spot.

Generally, a local station contracts with a national sales representative company, which represents the local station to the national advertiser. These national advertising representatives work in the nation's largest cities—New York, Los Angeles, Chicago, Detroit, San Francisco. Since the local station pays the rep a percentage of the sale as a commission, the rates for national advertising must be higher. The national sales representative is the only practical route for a local station to attract national spots, since it would be too costly for the local station in the middle-to-small market to send station personnel regularly to the major cities to deal with national advertisers.

Production Costs

Another rate differential involves the amount of work the station must put into producing the spot. A thirty-second spot with a station announcer reading a message is less expensive than a spot demanding full production—music, actors, and sound effects. Many local spots are produced by the local station, while most national spots are prepared by an advertising agency and are given fully prepared to the local station. The sales staff figures the production costs into the advertising contract. In setting a price, the station must consider the use of station facilities and equipment, of station personnel, of copyrighted music, of special talent, and special editing needs.

Each station sets its advertising policy on the basis of local demands. Some stations will accept only those ads that can be read by a station announcer and require no production; other stations accept only full production ads that make the advertising as entertaining as the program content. Some stations are building highly profitable sidelines by producing commercials. The necessary equipment is not prohibitively expensive and, by producing commercials, the station can charge more for its spots and offer the advertiser a unique service.

THE RADIO ADVERTISING MARKET

Commercial Growth

Since the 1960s, radio revenues have continued to grow. In recent years, commercial radio has earned approximately four billion dollars in revenues, which represents nearly 30 percent of total broadcast revenues and over $300 million in profits.[2] In 1966, radio's total revenues were only $872 million;[3] in just a few years that figure has increased by over three billion dollars. While some of the increases in total revenues can be attributed to inflation, a good portion of that figure represents real growth for commercial radio.

National Sales

One contributor to growth for radio is the national advertiser, who is again returning to radio. The Wrigley gum company, for example, began using radio for coast-to-coast advertising in 1972. Company officials claim that Wrigley uses radio for the following reasons:

1. The radio spots strengthen and extend the Wrigley television ads.
2. The radio spots, which can be aired on thousands of stations throughout the broadcast day, add momentum to the advertising campaign.
3. The radio spots reach consumers out of the home.
4. The radio spots reach young and ethnic consumers.
5. The radio spots provide tremendous frequency to the campaign across a wide variety of demographic groups.
6. The radio spots "drive home a musical message."[4]

Other companies have registered similar support of radio as a strong advertising medium. The Kraft Company, best known for its cheese products, dou-

2. *Broadcasting Yearbook,* 1983.
3. The National Association of Broadcasters.
4. *Broadcasting,* July 3, 1978, p. 42.

bled its radio advertising in recent years.[5] OXY-5, an acne medication, started using radio in 1975 as its only advertising medium. By 1977, retail sales of OXY-5 were up 40 percent.[6] The OXY-5 company, which had placed all of its advertising dollars into radio, attributed its sales success to that medium.

Advantages of Radio Advertising

There are a number of reasons why advertisers are attracted to radio. One is the low cost of most radio advertising. According to *Broadcasting Yearbook,* the average thirty-second prime time network television spot now costs approximately $100,000; a thirty-second television spot for a program with a low audience rating averages about $50,000. Radio spots cost from $750 or more in major markets to less than a dollar in small towns.[7]

Another reason why advertisers are attracted to radio is that the medium reaches a tremendous audience. The Radio Advertising Bureau (RAB) reported that radio reaches more adults in a day or a week than any other major daily advertising medium. According to RAB, radio reaches 95 percent of adults eighteen years of age during a week, while television reaches 90 percent and newspapers reach 84 percent.

Also, with radio, the advertiser can select a specific target audience. With various formats attracting specific demographic groups, the advertiser can purchase radio spots that are aimed at the target audience most likely to purchase the product. Radio is perhaps the best medium available for reaching specific groups of individuals in every local community across the nation.

For all of these reasons, radio advertising revenues have grown and are expected to continue to grow. Indeed, radio has great potential as an advertising medium, but creative sales and promotion staff members can help to realize that potential. With creative and dedicated sales personnel, radio could account for more than its current 30 percent of the total broadcast revenues.

RADIO RESEARCH

The Audience Survey

In order to sell station time to an advertiser, the sales staff must have specific information about the station's audience. This knowledge comes from research—predominantly survey research.

Most radio research is contracted by the local station to a research firm that

5. Ibid.
6. Ibid.
7. *Broadcasting Yearbook,* 1983.

specializes in radio survey research. Arbitron is the largest radio survey research company, with centers geographically located to serve most areas of the country—New York, Chicago, Atlanta, Los Angeles, San Francisco, Dallas, Washington. Other companies offer radio research services, but Arbitron has a firm hold on the market. Its methods of survey research are understood by the broadcasters and by the advertisers and have become industry standards.

The Survey Sample

Survey research is a method of gaining information about a large group of individuals by interviewing a small subset of the large group. The keys to good survey research are: 1) selecting a sample that is large enough to be representative of the entire group and 2) selecting a sample that represents the many interests of the larger group. In radio and television survey research, all social groups must be represented—minority groups, political conservatives and liberals, large families and small families, blue collar workers and professionals, housewives and employed women, and others.

Minority groups have argued that they are underrepresented in most television/radio surveys. Questions concerning the representation of minority groups in radio/television research have become so common that Arbitron issued a publication called *How Arbitron Radio Measures the Black Population*[8] to quiet some of these concerns. Despite publications like the one developed by Arbitron, there are still concerns that the survey research samples do not adequately represent all groups. Definitely, selecting the right sample is the first step to good survey research, and finding the "right" sample is no easy task.

Some research firms rely on telephone directories to select a random sample, but this method means that certain individuals will be excluded—those who do not have a telephone and those who have an unlisted number, for example. Frequently, this includes the poor, who may not have telephones, and the upper income groups who can afford to pay extra for an unlisted number.

Other methods for selecting a sample include using the zip code areas in a city to select representatives from each code area. The zip codes divide the city into manageable population areas, and frequently a zip code area identifies a homogeneous group that lives in that area. Also, sometimes zoning diagrams of a city, which are available from city and county officials, are used to divide the city into survey areas. Whatever the method for selecting a sample, the researcher must be able to prove that the method adequately represents all the groups who live in the city or in the station's market area.

Selecting a sample for radio research is frequently more difficult than selecting a sample for television research. The reason is that in most locations

8. *How Arbitron Radio Measures the Black Population,* Arbitron Research Bureau, 1350 Avenue of the Americas, New York, N.Y. 10019.

there are fewer TV stations than there are radio stations, and at any given time more people are likely to be watching one of the television stations than listening to one of the radio stations. Radio researchers have explained the problem this way.

> Radio ratings are generally less definite than television ratings, for several reasons. The radio audience is more fragmented, with more stations to be rated. A sample size of 1,000 is more meaningful for five television stations than for thirty or forty radio stations. For example, if 1,000 units were sampled at a given time of an evening and twenty percent were listening to radio, only 200 of the people contacted would be listening to forty radio stations in the market. That leaves only two people per rating point. The percentage using radio at some times of the day can be quite low.[9]

Research Methods Comparison

Each research company has defined its methods of gaining information about the listening audience. One frequently used by Arbitron is the research diary. The diary is a collection of sheets of paper for each day of the week that provides spaces for the following information: 1) specific times of the day the individual listened to the radio, 2) the station call letters of the station listened to, 3) whether the station was AM or FM, 4) where the individual listened to the radio—in the home or outside of the home. Exhibit 8.2 shows an Arbitron research diary.

When an Arbitron diary is placed in a home, each member of the household over twelve years old is given a diary. This enables the researcher to gain information from a variety of age groups. The family is instructed to mail in each day's diary as it is completed. The process of daily mailing reminds the family to record the day's radio listening, and the research company gets fairly rapid results.

Another method of radio research is the telephone survey. The telephone interviewer calls a home and asks: "Are you listening to the radio?" "Is anyone in your household listening to the radio?" If the answer is "yes" to either of these questions, the next question is "What station or stations are being listened to?"

One obvious disadvantage of telephone surveys is the above-mentioned problem that not everyone is listed in the telephone directory. Another problem is that the information that can be obtained over the telephone is limited. Also, one family member speaks for the entire family.

Door-to-door survey research is another method used to gather information about radio listenership. This method avoids one problem inherent in telephone surveys—it includes some families that would be excluded by the

9. Joseph S. Johnson and Kenneth K. Jones, *Modern Radio Station Practices,* 2d ed. (Belmont, Calif.: Wadsworth, 1978), p. 216.

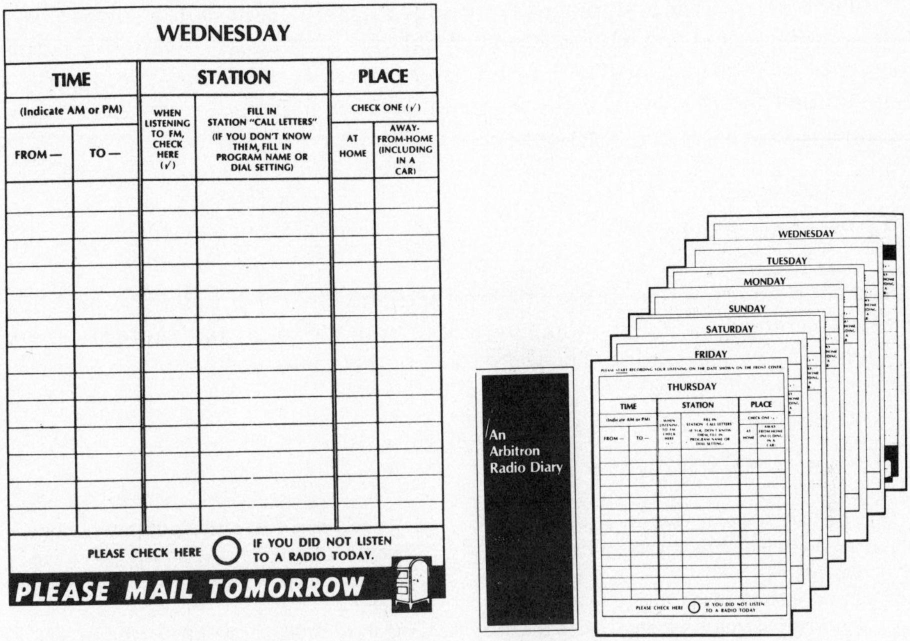

Exhibit 8.2.   The Arbitron Radio Diary
The Arbitron Radio Diary runs from Thursday through Wednesday. It asks the listener to record stations listened to, time of day when listening occurred, FM or AM station, and whether the listening was done in the home or outside the home. (Photo courtesy of Arbitron.)

lack of a telephone. In many door-to-door surveys a *cluster* method is used in which the interviewer may find the survey addresses in the phone directory but starts the first interview at the house next door to the one selected in the telephone book. This procedure provides an opportunity for including some families that may not have phones. The sample is called a *cluster* because several houses in the same neighborhood are included in the survey. The reason for surveying in a cluster is purely economic; since the interviewer is in the neighborhood, it is most economical to interview several families before traveling to a new location.

The interviewer usually asks the person who answers the door for specific information on household radio listening. If possible, the interviewer talks with each member of the household over twelve years of age. The interviewer asks the respondents what radio station(s) each listened to that day, what time the listening occurred, and whether the listening was at home or outside the home. The interviewer may also get information about the household size, educational background, household income, and other data.

There are both advantages and disadvantages to each method of research.

Exhibit 8.3 shows the advantages and disadvantages of radio survey research by diary, telephone, and door-to-door. What the broadcaster wants from this research is information that can be used in making decisions about station operations and practices; if the research data are inaccurate, the station personnel may make poor decisions.

Ratings and Shares

Audience survey research provides the station with several kinds of information. The *rating* is the percentage of people from the total market area who are listening to a given station. The *share* is the percentage of all people listening to

DIARIES

ADVANTAGES	DISADVANTAGES
1. The diary provides a day-by-day account of radio listening over a substantial time period.	1. Some individuals do not accurately report radio listening in the diaries.
2. The respondent isn't forced to recall information from past days, since the diary is completed each day.	2. Some individuals do not regularly mail the diaries.
3. Since the diary can be a constant companion for recording radio listening, more complex information may be obtained from each respondent.	3. Only 65 to 70 percent of the returned diaries are useful.

TELEPHONE SURVEYS

ADVANTAGES	DISADVANTAGES
1. The results can be rapidly tabulated.	1. The survey is limited to those individuals with telephone numbers listed in local directories.
2. The interviewee isn't forced to rely on recall, since the interviewer asks only about radio listening at the time of the phone call.	2. The sample must be fairly large to insure meaningful results.
3. Telephone interviews are fairly inexpensive.	3. Only a limited amount of information may be obtained over the telephone.

DOOR-TO-DOOR SURVEYS

ADVANTAGES	DISADVANTAGES
1. Detailed information may be obtained from the respondents.	1. Requires extensive personnel time for the interviews.
2. Avoids the problem of selecting a sample with telephone numbers.	2. Is a fairly expensive survey method.
	3. Takes longer to gather data than telephone survey.

Exhibit 8.3.   Advantages and Disadvantages
of Radio Survey Methods

radio who are tuned to the particular station being surveyed. If 40 percent of all people in a given community are listening to radio and 20 percent of people in the community are listening to one station, the station's rating is 20 percent and its share is 50 percent.

Two other pieces of information provided by the research companies are: quarter-hour average listeners and cumulative ratings. The quarter-hour figure tells the station how many persons were tuned in for at least five minutes during a fifteen-minute survey period. The quarter-hour figure can tell the station when listeners are most likely to tune in.

The cumulative figure is the total number of unduplicated listeners who tune in during a survey period. If a listener tunes into a station from 7:00–8:00 AM and again from 2:00–3:00 PM, the listener would be counted only once in a daily cumulative survey. The cumulative rating provides the station with information about the total number of individuals in a market that the station is reaching. The "cume" can also be divided by age and sex for more specific information on the total reach of the station.

The research companies publish rating books that provide information about all the radio listening in a given market area. Stations that subscribe to the rating service regularly receive copies of the books. Arbitron divides the broadcast day into time segments and provides detailed information about each station's listeners at the various times—6:00 AM–10:00 AM; 10:00 AM–3:00 PM; 3:00 PM–7:00 PM; 7:00 PM–Midnight—as well as other time divisions. Arbitron also provides information on listening patterns by age; men 18+, men 18–34, men 18–49, men 25–49, men 25–54, men 35–64 (the same age categories for women are provided). Teenagers form a special section in the rating book. Exhibit 8.4 shows sample pages from the Arbitron reports.

Use of Radio Research

How can the station use the information provided by ratings, shares, quarter hours, and cumes? The following list is an example:

1. To aid the station in adjusting its format. If one format draws a large audience, the station may decide to expand that format or to eliminate a format or to seek a specialized audience through a special format.
2. To attract advertisers to sponsor a radio program or to purchase spots for particular times during the broadcast day.
3. To study station listenership trends over time.
4. To aid station personnel in making decisions concerning format and sales and promotion campaigns.
5. To compare one station with another—formats, listenership, or audience mix at various times during the day.
6. To prepare demographic information about station listenership to be used by station personnel.

Trends

Audience Trends

The Trend Section gives a flash report on what's happening in a market on the basis of broad demographics, and is the first section with audience estimates in the report. Trends allow the report user to observe what has been the relative standing among stations over a period of time.

The estimates are:
Average Persons Share —
Metro Survey Area

The demographics are:
Total Persons 12+
Men 18+
Women 18+
Teens 12-17

The day-parts are:
Mon-Sun 6AM-Mid
Mon-Fri 6-10AM
Mon-Fri 10AM-3PM
Mon-Fri 3-7PM
Mon-Fri 7PM-Mid

Average Share Trends—Metro Survey Area

APRIL/MAY 1977

TOTAL PERSONS 12+

MON-SUN 6:00 AM-MID

STATION CALL LETTERS	A/M '76	O/N '76	J/F '77	A/M '77
WAAA	1.1	1.0	.6	1.0
WBBB	4.5	4.9	6.2	5.4
WCCC	3.7	5.1	2.3	3.1
WDDD	3.7	3.2	3.3	2.8
WEEE	2.1	2.5	1.2	2.2
WFFF	2.1	2.7	2.6	2.2
WGGG	4.1	3.6	3.8	3.9
WHHH	**	**	**	.6
METRO TOTALS	16.1	16.3	15.2	16.4

MON-FRI 6:00 AM-10:00 AM

STATION CALL LETTERS	A/M '76	O/N '76	J/F '77	A/M '77
WAAA	.6	.8	.8	.8
WBBB	3.4	3.8	3.7	3.9
WCCC	4.9	3.4	3.0	3.4
WDDD	3.1	2.6	3.0	2.7
WEEE	2.0	2.7	1.1	2.6
WFFF	1.5	3.4	1.7	2.2
WGGG	6.6	5.1	6.4	5.7
WHHH	**	**	**	.3
METRO TOTALS	21.5	20.0	21.6	22.4

MON-FRI 10:00 AM-3:00 PM

STATION CALL LETTERS	A/M '76	O/N '76	J/F '77	A/M '77
WAAA	1.4	1.0	1.0	1.2
WBBB	5.7	6.4	8.2	6.9
WCCC	4.4	3.3	3.0	3.6
WDDD	3.2	3.4	2.9	2.7
WEEE	1.7	1.7	1.3	2.9
WFFF	1.9	2.3	2.5	2.1
WGGG	5.3	3.2	2.7	3.4
METRO TOTALS	16.1	20.6	18.4	16.7

MON-FRI 3:00 PM-7:00 PM

STATION CALL LETTERS	A/M '76	O/N '76	J/F '77	A/M '77
WAAA	1.2	1.1	.7	1.4
WBBB	4.7	5.1	6.8	5.7
WCCC	3.6	3.0	2.4	3.1
WDDD	3.7	3.2	3.4	2.7
WEEE	2.0	1.6	.8	1.5
WFFF	1.8	2.2	2.9	2.7
WGGG	3.7	4.1	3.3	3.5
METRO TOTALS	18.2	18.2	17.4	18.7

MON-FRI 7:00 PM-MID

STATION CALL LETTERS	A/M '76	O/N '76	J/F '77	A/M '77
WAAA	1.8	1.5	.9	.9
WBBB	4.0	4.2	5.0	4.9
WCCC	1.6	1.8	2.2	1.8
WDDD	4.6	4.0	3.4	3.6
WEEE	3.0	1.7	.7	.6
WFFF	2.9	2.5	3.8	3.6
WGGG	4.1	3.4	1.2	4.5
WHHH	**	**	.7	.3
METRO TOTALS	18.2	16.6	15.0	16.1

Footnote Symbols: (-) means audience estimates adjusted for actual broadcast schedule (**) Station not reported or reported under different call letters this survey (+) means AM-FM Combination was not simulcast for complete time period

ARBITRON

Metro totals represent market listening levels expressed as a rating.

168

Hour-by-Hour Estimates

The Monday-Friday Hour-by-Hour estimates, shown from 5AM to 1AM, pinpoint station performance in periods larger and smaller than standard day-parts.

The demographics are:
- Total Persons 12+
- Men 18+, 18-34, 18-49, 25-49, 25-44, 35-44, 45-54
- Women 18+, 18-34, 18-49, 25-49, 25-44, 35-44, 45-54
- Teens 12-17

The estimates are:
- Average Persons, Metro and Total Survey Area
- Shares, Metro Survey Area (Persons 12+, Men 18+, Women 18+ and Teens)

Hour By Hour Avgs

METRO SHARES

MONDAY-FRIDAY 6:00AM-7:00AM

STATION CALL LETTERS	TOT. 12+ %	MEN 18+ %	WM. 18+ %	TNS 12-17 %
WAAA				
WBBB				
WCCC				
WDDD				
WEEE				
WFFF				
WGGG				
WHHH				
WIII				
WJJJ				

Average Quarter-Hour Listening Estimates

AVERAGE PERSONS—METRO SURVEY AREA, IN HUNDREDS

(MEN: 18-34, 18-49, 25-49, 25-44, 35-44, 45-54, 18+; WOMEN: 18-34, 18-49, 25-49, 25-44, 35-44, 45-54, 18+; TNS 12-17)

AVERAGE PERSONS—TOTAL SURVEY AREA, IN HUNDREDS

(MEN: 18-34, 18-49, 25-49, 25-44, 35-44, 45-54, 18+; WOMEN: 18-34, 18-49, 25-49, 25-44, 35-44, 45-54, 18+; TNS 12-17)

TOTAL LISTENING IN METRO SURVEY AREA

ARBITRON

Footnote Symbols: (*) means audience estimates adjusted for actual broadcast schedule (+) means AM-FM Combination was not simulcast for complete time period.

Exhibit 8.4.   The Arbitron Radio Survey Reports

A radio station that subscribes to Arbitron receives a ratings book after the market survey of listening audience has been completed. The ratings book contains information about the numbers of males, females, and teens who listen to the various stations in the market. The information is provided by times of the day—6:00 AM–Midnight, 6:00 AM–10:00 AM, 6:00 AM–7:00 PM, and 3:00 PM–7:00 PM. (Photo courtesy of Arbitron.)

169

Sometimes stations use the ratings information to argue that their station is "number one" in the market. This may be true, but it is not always easy to prove from the ratings. In the same market area, one station may be number one in terms of teenage listeners; another may attract the most women eighteen to forty-nine years of age. One station may attract more listeners during its regular newscasts than any other station, but it may trail other stations during its music format. One station may have the highest weekly cume rating and trail other stations during certain times during the day. The important point to remember is that the ratings provide a great deal of information, but that information must be properly interpreted to be meaningful. When trying to determine which station is number one in a given market, many factors must be considered.

Surveys as Estimates

When looking at the ratings, some station personnel panic if the station drops in the ratings by a fraction of a point. Actually, the ratings are only *estimates* of audience size and are not hard figures. There are two kinds of errors in survey research—statistical errors and sampling errors. The statistical errors can be calculated and are reported in the data. If a rating is 8 percent and has a standard error of ± .6 this means that the true audience rating has a high probability of being in the range of 7.4 to 8.6 percent. The statistical error can be calculated and is used in interpreting the findings, but the sampling error cannot be calculated. If the researcher excluded from the sample whole segments of the population, then the results of the research would be misleading. This is the reason for concern about a representative sample and the reason for using the sampling method that is most likely to represent the entire community. Only repeated studies using different samples can yield information on the sampling error.

Since there is a statistical error associated with each rating, there is usually no reason for concern if the station slips a fraction of a point. A station with a rating of 10.5 may have the same audience as a station with a rating of 9.8. Station management does, however, get worried whenever the ratings seem to slip, regardless of how small the change.

The problem is that the station manager knows that advertisers do rely on the ratings, and thus the ratings become all important. In radio and television broadcasting, ratings have become the foundation upon which all programming decisions are made and upon which the station promotes itself to the public and to the advertiser. Through the 1980s and into the 1990s, all signs indicate that ratings will be king.

Individuals who use ratings to make decisions in radio should always be cautious of the findings until questions have been answered regarding the research methods and until the findings have been placed in proper perspective. Some good questions to ask about any broadcast research are:

1.   What methods were used in the research?
2.   How was the sample drawn?
3.   What questions were asked?
4.   How were the findings statistically interpreted?
5.   How can the findings be used to aid the station in its decision making?

Small Station Research

Many small stations do not fall within the regular survey areas of Arbitron or other large research companies. Also, the small stations may not be able to afford the regular costs of subscribing to a survey research company. Even the small station, however, needs information about its listening audience.

A small station may conduct its own research. It can conduct a telephone survey of its market area or a door-to-door survey. A self-survey will provide the station personnel with information upon which to base decision making. A self-survey will also provide some information about the station audience to the potential advertiser.

Another method for obtaining information on the listening audience is to hire an independent research firm for a one-time survey. This is considerably less expensive than a regular subscription to a survey research company. The information the station gathers in the one-time survey can be used for several years. Many research firms offer special services to the individual radio station. The research firms and their service options are the next topics for discussion.

Research Firms

If the radio station is willing to pay the price, the station can get almost any kind of information about its listeners and about its programming or decision-making processes. *Broadcasting Yearbook* lists numerous research firms that offer a wide variety of services—pretesting radio commercials, radio market research, programming research, advertising impact studies, feasibility studies on any project the station may want investigated, station "image" research, news programming research, rate card studies, and others.

The more information the station seeks, the more the research will cost. Station management must decide how much research it needs to make meaningful decisions on programming and sales and promotion. There is no question that the 1980s and 1990s will be the heyday for research firms. Air personalities will be added and dropped on the basis of ratings, and promotion campaigns will be designed around such ratings. The job for the radio broadcaster is to know what the research findings actually mean and to understand how to apply the findings effectively to station operations.

PROMOTION

Building the Audience

In addition to using research to aid the station's sales efforts, research is also used to aid promotion efforts. Promotions are the various campaigns that the station uses to increase its listening audience.

Most radio managers base the station's promotion efforts on the market competition. In a large market area, New York or Los Angeles, for example, where thirty to forty stations are in competition with each other, promotion efforts must be almost constant in order to stand out in the crowded spectrum. In a small town in Montana, where the community can receive only two or three stations, the promotion efforts are much less intense. The first question any station manager must ask about promotion is: how much promotion is necessary for this particular market?

The promotion efforts are designed to increase audience size and thus increase revenues. If the station management decides that promotion efforts are needed, a series of questions must be posed and answered before the promotional personnel launch a promotion campaign.

Planning a Promotion Campaign

The first question is: what specifically does the station need to accomplish in a promotion campaign? Attract more teens for specific hours? Appeal to more women eighteen to forty-nine years of age? Attract more men with incomes above a certain level? Specifically, the promotion personnel must understand what type of audience building is needed.

The second question that must be answered is: how will the promotion efforts be accomplished? What type of campaign will be used? The station has a wide variety of available options:

1. Launch a station contest.
2. Advertise the programs that need larger audiences during times when members of the target audience may be listening.
3. Advertise the station by using other media—television, newspapers, local newsletters, billboards.
4. Advertise the station by using banners on local buses or placards in taxis.
5. Advertise the station by sponsoring local events—concerts, civic activities, sports events, or picnics.

The staff must decide exactly what type of audience it wants to build and then must plan a promotion designed especially to attract those individuals. Exhibit 8.5 shows a newspaper promotion designed to attract a larger teen and young adult audience on Saturdays.

**EXCLUSIVELY
IN CENTRAL IOWA**

THE WEEKLY

TOP THIRTY
THE RADIO SHOW

**THE WEEK'S BEST HIT MUSIC
EVERY SATURDAY
ONE TILL FOUR P.M.**

ON

Q 107 KCCQ, AMES
AN IOWA RADIO NETWORK STATION

Exhibit 8.5.   A Newspaper Promotion for Radio
This is an example of a promotion designed to attract a young adult audience and
was part of a promotion campaign conducted in a newspaper. (Photo courtesy of
KCCQ, Ames, Iowa.)

A third question that must be asked is: how much money should the station
invest in the promotion? The staff must do some careful calculations to determine how much increased revenues can be generated by increasing audience
size. If the campaign costs the station more revenue than an increased audience will generate, can the promotion be justified?

Under some circumstances, the station may decide that the promotion is
valuable even if the audience size does not increase. The station may want to
maintain its present audience, rather than expand. The promotion campaign
would be designed to assure listeners that they should continue listening. The
station may want to generate community "good will" for the station; no dollar
value can be placed on the good reputation that a station earns in its community. In this case, the station may want to support and sponsor charity drives or
other community events. In another case, the station may just want to keep its
name before the public.

There are also other times when a station may want to sponsor a promotion
campaign. When a new owner purchases a station, the new owner may want to
introduce himself or herself to the community. When the station adds a new
format or changes an old one, a promotion campaign would definitely be in
order. A station may also want to announce to the public new air personalities
or new station policies. Whatever the purpose of a campaign, the costs must be
weighed against the benefits to the station.

The Campaign in Action

Once the campaign has been decided upon, personnel must be assigned to specific tasks. Who will do what when? Without individual tasks and specific deadlines, nothing will be accomplished. The promotion department needs good leadership—a leader who can inspire and organize the staff.

The specific assignments and deadlines must be regularly discussed and evaluated. One good way of reminding the staff of tasks and deadlines is to develop a chart or poster showing the specific tasks with deadline dates. As each task is completed, the poster is revised to indicate that the particular job has been completed and that the staff is moving toward accomplishing the next goal. The visual reminder for the staff is also a good form of peer pressure for the apathetic worker who has a tendency to put tasks off to the last minute.

If the promotion involves using newspaper ads to promote the station, who will design the ads and when will they run? If the promotion involves designing new materials for distribution to potential advertisers, who will design the material? How will the materials be distributed? If a contest is designed to attract new listeners, what station personnel will participate in the contest? Who will be responsible for promoting the contest? The point is that every phase of a promotion campaign should be carefully developed and assigned to a specific person to be completed by a designated date.

Special Services for Radio Stations

If a station does not have the promotion staff or expertise to design a promotion campaign, the station can purchase promotion services from a variety of radio service representatives. Many format service representatives furnish stations with musical jingles for station identifications, and many offer subscribing stations regular audience surveys and advice on attracting particular types of audiences.

Broadcasting Yearbook devotes an entire section to public relations and promotion companies that will design a promotion campaign for a station. Also, there are consulting firms that will do research to help the station determine if a promotion campaign is in order and, if so, the exact dimensions the campaign should take.

Whether or not a station decides to use any of the available services, a good broadcaster should know what is available. Any student with career aspirations in radio broadcasting should spend some time studying the available services to broadcasters, including advertising agencies, program services, station representatives, consultants, public relations and promotion services, research services, production services, employment services, and other industry-associated services.

Working in the sales and promotion department is very frequently the route to commercial radio station management. Since sales and promotion are the backbone of station operations, a manager should thoroughly understand the workings of this vital department.

BIBLIOGRAPHY

Bittner, John. *Professional Broadcasting: A Brief Introduction.* Englewood Cliffs, N.J.: Prentice-Hall, 1981.

Broadcasting. Issued weekly throughout the year.

Broadcasting Yearbook, 1983.

Coleman, Howard W. *Case Studies in Broadcasting Management.* New York: Hastings House, 1975.

Johnson, Joseph S., and Jones, Kenneth K. *Modern Radio Station Practices.* 2d ed. Belmont, Calif.: Wadsworth, 1978.

Quaal, Ward L., and Brown, James A. *Broadcast Management.* 2d ed. New York: Hastings House, 1976.

Robb, Scott H. *Television/Radio Age Communication Coursebook.* New York: Communication Research Institute, 1978.

QUESTIONS FOR REVIEW AND DISCUSSION

1. Define: a) CPM, b) regular sponsor, c) scatter spots, d) national rep.
2. In radio advertising, why is the quantity spot beneficial to the advertiser and to the station?
3. What factors must be considered in setting advertising rates?
4. Why is radio a good medium for the advertiser?
5. What are the keys to good survey research?
6. Why is selecting a survey sample for radio audiences more difficult than selecting a survey sample for television audiences?
7. What survey data recording method is used by Arbitron?
8. Define: a) rating; b) share; c) cume.
9. What are the advantages and disadvantages of the following survey methods? a) diaries, b) telephone surveys, c) door-to-door surveys.
10. How can a station use the information provided by ratings, shares, and cumes?
11. Can the ratings be used to prove a station is number one in a given market?
12. What are the two kinds of errors inherent in survey research?
13. How do advertisers use radio audience survey research?
14. What questions should always be asked about broadcast survey research?
15. What kinds of research services are available to the radio station?
16. What are the responsibilities of the station promotion department?
17. What is the rationale of most station promotion efforts?

18.   Once a station has decided to invest in a promotion campaign, what three questions should be asked and answered?

19.   Under what circumstances might a station invest in a promotion campaign not directed at audience building?

20.   What types of options are available for radio promotion campaigns?

21.   What can the promotion manager do to insure a successful promotion campaign?

22.   What types of promotional services may a station purchase from broadcast service companies?

23.   Why are sales and promotions the backbone of radio station operations?

Chapter Nine

GOVERNMENT AND OTHER REGULATORS OF RADIO

These columns have sagged beneath the weight of protests against the chipping away at radio's freedom. For 15 years this has gone on—since the first issues of this publication. Betimes we have been accused of calamity-howling, of protesting too much.

The answer is evident in what has happened. Regulatory measures never dreamed of have been invoked by the Federal Radio Commission and its successor FCC. The thought implicit in the Radio Act of 1927, and carried over into the Communications Act of 1934, that the FCC is restricted to regulations of the physical aspects of broadcasting, is almost as extinct as the dodo.

"How Far Can the FCC Go in
Regulating Broadcasting?"
Broadcasting, January 28, 1946

Radio is regulated directly and indirectly by a variety of forces—some within the federal government, others within the industry itself, and, indirectly, others within the general community. In the United States, our communications media are supposed to function free of governmental restraint, creating a forum for the dissemination of all kinds of information. Our communications media are among the freest on the globe, but the broadcasting media *are* regulated by federal agencies. These regulations developed because of the uniqueness of the broadcast media. This chapter discusses the various regulations that have been imposed on the radio industry and explores the rationale for regulating broadcasting as well as the exact nature of the various regulations.

FEDERAL INTERVENTION

Regulating Broadcasting

As has been discussed in earlier chapters, regulation of the broadcasting industry was absolutely essential for the development and growth of the industry. Broadcasters in fact encouraged the federal government to involve itself in regulating radio. The reason that many broadcasters wanted regulation was to have their transmissions heard, free of interference. Broadcasters were using whatever frequencies they arbitrarily decided to use, and the general public was the big loser, since the public had difficulty receiving a clear radio signal among the many competing radio waves.

Finally, in 1927 the first comprehensive broadcasting legislation was adopted by the federal government—the Radio Act of 1927. The act established a five-person Radio Commission that was given responsibility for classifying stations and for defining the exact nature of service that each class was to render. The commission also maintained the authority to issue broadcast licenses to radio operators and to stations and to refuse to issue licenses or to revoke licenses of those individuals or stations not functioning within the provisions of the act.

The Radio Act of 1927 was revised to become the Communications Act of 1934, which is the federal regulatory document that governs broadcast media even today. Both the Radio Act of 1927 and the Communications Act of 1934 clearly established the principle that the general public should benefit from the commercial use of the airways.

A rewrite of the Communications Act may be undertaken in the near future, and it seems warranted because new technology has made the 1934 act obsolete. Cable, satellites, fiber optics, digital electronics, and even television are not covered by the 1934 act. Congress has considered rewriting the act but any rewrite is sure to be highly controversial. Anyone interested in broadcasting should follow the trade publications for important developments as the 1934 act continues to be reviewed.

The Scarcity Doctrine

The justification for imposing regulations on broadcast media that are not imposed on any other media has been called the "doctrine of scarcity." The doctrine argues that there are a limited number of frequencies available for broadcasting, which means that not everyone can engage in commercial radio transmission. This situation is not paralleled in the print industry where anyone with access to a typewriter and some facilities for printing or copying can publish written documents—a newsletter, a newspaper, a magazine, or tract. The doctrine argues that because of the scarcity of broadcast frequencies, the airwaves should be considered as belonging to the general public and should be used to serve the general public.

This belief that radio and television communications are unique and that they should be regulated has been sanctioned not only by Congress, but also by the courts. In 1966, Justice Warren Burger explained:

> A broadcaster has much in common with a newspaper publisher, but he is not in the same category in terms of public obligations imposed by law. A broadcaster seeks and is granted the free and exclusive use of a limited and valuable part of the public domain; when he accepts that franchise, it is burdened by enforceable public obligations. A newspaper can be operated at the whim or caprice of its owner; a broadcast station cannot.[1]

There is some question as to whether or not regulating a communication industry is in direct conflict with the First Amendment to our Constitution which states that: "Congress shall make no laws . . . abridging the freedom of speech, or of the press. . . ." The federal courts have examined this issue and have concluded that if regulating radio and television is in the public interest, there is no denial of freedom of speech. In 1943, the Supreme Court concluded:

> Unlike other media of expression, radio inherently is not available to all. That is its unique characteristic; and that is why, unlike other modes of expression, it is subject to government regulations.[2]

Some would argue that the "doctrine of scarcity" no longer applies to the broadcast media. With the addition of cable broadcasting systems, many new stations have been added to the TV and radio dials. Also, satellites promise more programming variety for broadcasting consumers. Public radio and television have introduced new programming formats into most communities. Has technology indeed made the "doctrine of scarcity" obsolete? This question will come forward again and again for debate throughout the 1980s and the 1990s.

Since radio transmissions cross state boundaries, radio was classified as interstate commerce and, according to the Constitution, Congress maintains the right to regulate interstate commerce. Congress had the responsibility for initiating a body of laws to regulate the broadcasting industry; that body of laws became the Communications Act of 1934.

The founding principle of the 1934 act is that a broadcast license should be granted "in the public interest, convenience, and necessity." This phrase appears often in the act and is used explicitly in the section that details the conditions for granting a station license:

1. *Office of Communication of the United Church of Christ v. FCC,* 359, 2d, 994 (D.C. Cir. 1966).
2. *National Broadcasting Co. v. United States, Columbia Broadcasting Systems, Inc. v. Same,* 319 U.S. 190, 87 L. Ed. 1344, 63 S. Ct. 997 (1943).

> The Commission, if public convenience, interest or necessity will be served thereby, subject to the Act, shall grant to any applicant therefore a station license provided for by this Act.[3]

While there is absolutely no question that Congress intended for radio to be regulated, there is also no question that Congress intended radio to be free from censorship. The Communications Act was written with few restraints on program content. In fact, in the first part of the act, Congress makes clear its commitment not to censor radio programming.

> Nothing in this Act shall be understood or construed to give the Commission the power of censorship over radio communications or signals transmitted by any radio station, and no regulation or condition shall be promulgated or fixed by the Commission which shall interfere with the right of free speech by means of radio communications.[4]

Commercial broadcasting, then, was to be considered a public commodity franchised to individuals who, in the process of earning a profit from broadcasting, would also serve the public good. To insure that the tenets of the Communications Act were enforced, Congress created the Federal Communications Commission, charged with enforcing the act.

The FCC

The Federal Communications Commission (FCC) is a seven-member commission appointed by the president of the United States with the approval of the Senate. The Senate holds hearings on any individual who is nominated to serve on the FCC. The hearings provide an opportunity for individuals who support or oppose the nominee to speak out. These proceedings guarantee that each FCC commissioner has been thoroughly scrutinized before being finally appointed to the commission. One of the seven members is appointed by the president to serve as chairman of the commission. To avoid stacking the commission with political appointees, no more than four of the seven commissioners may be from the same political party, and each appointee serves a seven-year term.

Like any agency of the federal government, the FCC is a bureaucracy with many facets. Regarding commercial broadcasting, the FCC's primary responsibilities includes:

Classifying stations
Prescribing the nature of service to be rendered by each class
Assigning frequencies

3.   Section 307a, *Communications Act of 1934.*
4.   *Communications Act of 1934*, Title III, Part I, Section 326.

Determining locations for the different types of stations

Regulating broadcast equipment usage

Prescribing qualifications for broadcast stations and operators

Publishing necessary information to alert stations to any changes in the regulations

Monitoring station operations to insure compliance with FCC rules and regulations

The Communications Act and the entire body of federal regulations and laws are extremely complex. Most stations employ both a local attorney and an attorney based in Washington, D.C., to insure that the station is in compliance with all the regulations. Because of the complexity of the regulatory process, the discussion of regulations in this chapter is not complete but is rather a survey of some of the basic broadcasting regulations. The first items for discussion are regulations that directly affect programming.

SECTION 315: THE EQUAL TIME RULE

One area of programming that is uniquely regulated by the Communications Act of 1934 is political broadcasting. Section 315 of the act, which is commonly called the "equal time clause" states:

> If any licensee shall permit any person who is a legally qualified candidate for any public office to use a broadcast station, he shall afford equal opportunity to all other such candidates for that office in the use of the broadcast station.

This means that if a broadcaster invites one candidate, who is running for mayor, to the station to be a guest on the local public affairs program, all other mayoral candidates must also be afforded equal time on the air. If one candidate is *given* fifteen minutes air time, every other candidate for that particular office must also be offered fifteen minutes of free air time.

Few regulations apply directly to programming, but the equal time rule is a rigid standard that does. Congress felt justified in intervening in the presentation of political broadcasts, since citizens should receive information about all candidates who are running for a particular office.

In addition to the equality of actual air time, the rule also states that stations must sell air time to all political candidates, both local and national, at the lowest available rates. This means that the station must give any special discounts and rate advantages to all candidates equally.

The equal time rule does allow a station to air news in which the candidate may be a participant. The equal time rule does not apply to a candidate's appearance on:

1.   A bona fide newscast
2.   A bona fide news interview

3. A bona fide news documentary (if the appearance of the candidate is incidental to the presentation of the subject covered by the documentary)
4. On the spot coverage of bona fide news events

These exemptions allow a station to carry news in which a candidate may be directly or indirectly involved. The difficulty is in deciding what constitutes bona fide news. In deciding whether or not a program or a program segment is bona fide news, the FCC considers such factors as the format, the nature and the content of the program, and the events covered in the program.

Some exemptions have been made to the equal time rule. In 1960, Congress temporarily suspended part of Section 315 to allow the Kennedy-Nixon debates. In 1975, the FCC declared that a station may now carry debates between two major party candidates without minor party candidates participating in the debates. The intent of the equal time rule has always been to provide the public with exposure to political candidates.

Section 315 or the equal time rule is being reviewed by the FCC. The commission may recommend that the provision be repealed outright or that the equal time provision conflicts with First Amendment rights of broadcasters. Any discussion of the repeal of Section 315 or of major revisions in this section are sure to be controversial.[5]

THE FAIRNESS DOCTRINE COVERING CONTROVERSIAL ISSUES

Because of the scarcity of broadcast freqencies, Congress and the FCC have always supported the position that the airwaves should be used to inform the American public of important public issues. The FCC adopted the fairness doctrine to promote the exchange of controversial ideas in the public's airwaves.

What the doctrine says, in essence, is that stations should *encourage* the exchange of ideas on important and controversial issues. The fairness doctrine does not require stations to provide *equal time* for the exposure of two or more points of view on a controversial issue. But the doctrine does make it clear that opposing sides should be heard and that the various positions should be treated in approximately the same way.

This means that if a station airs one side of a controversial issue during morning drive time when radio listenership is high, it is not fair to air the opposing side at 4:00 AM when listenership is very low. It is not fair to give one side fifteen minutes and another side only two minutes. It may be fair, however, to

5. Since 1945, broadcasters have been allowed to air political debates between major party candidates as long as a third party sponsored the debate. In early November 1983, the FCC ruled that broadcasters are now free to sponsor debates on their own without triggering equal time demands. See *Broadcasting,* November 14, 1983, p. 33.

provide one side with fifteen minutes of coverage during afternoon drive time and the other side with five minutes of drive time for three consecutive mornings. It may be fair for one side to be presented at 7:30 AM and the other side at 5:10 PM, since both times have large audiences.

In the determination of fairness, it is a station's *total programming* that is considered. The key to fairness is approximation—a close approximation in total time should be provided for the various sides. Also, similar types of air time should be provided, and the issues should be treated in a roughly similar manner.

Other media—books, magazines, newspapers, films—are not held to any standards of fairness. Broadcasters have argued that the fairness doctrine violates their First Amendment rights, but the Supreme Court held that:

> There is nothing in the First Amendment which prevents the Government from requiring a licensee to share his frequency with others and to conduct himself as a proxy or fiduciary with obligations to present those views and voices which are representative of his community and which otherwise, by necessity, would be barred from the airwaves.[6]

Broadcasters argue that the fairness doctrine has just the opposite impact from what it was intended to have. They argue that the doctrine hinders rather than encourages the presentation of controversial ideas. Since the fairness doctrine applies to *all* programming, station managers argue that they fear airing any controversial issues because they must then seek opposing opinions. They argue that the doctrine has a chilling effect on the free flow of controversial ideas in the broadcast media.

The doctrine also raises some interesting questions. Do unpopular opinions have just as much right to be heard as popular ones? Are broadcast advertisements covered by the doctrine? How actively must the station management pursue the presentation of controversial issues? Answers to these and other questions are frequently vague. Thus, compliance with the doctrine is sometimes difficult for the broadcaster.

In recent years, the fairness doctrine has come under heavy criticism—even from FCC commissioners. The FCC may seek to eliminate or modify the doctrine. The justification for this is that recent technological developments have made the doctrine obsolete. The number of radio and TV stations on the air has increased dramatically; cable systems have added many new radio and TV channels to local markets; numerous satellite radio and TV networks will be available to local communities. Some commissioners are prepared to argue that with the new broadcast services, competition in the marketplace will insure the coverage of controversial ideas.

6.   *Red Lion Broadcasting Co. v. FCC,* 395 U.S. 367 89 S. Ct. 1974, 23 L. Ed. 2d 371 (1969).

The fairness doctrine is an important feature of American broadcasting. Any discussion of its modification or elimination will be highly controversial and will include discussions from members of Congress, the FCC, the broadcasting industry, consumer groups, and the general public.

THE PERSONAL ATTACK RULE

Another facet of the fairness doctrine is the "personal attack rule." In the airing of controversial issues, occasionally a person or a group may be denigrated. The FCC made a special provision in the case of a person or group being attacked on the air. If an attack is made on "the honesty, character, integrity, or like personal qualities" of an identified person or group, the station must: 1) send a notice to the attacked person or group within one week, 2) send a tape or transcript of the program or segment in which the attack occurred, and 3) offer time for the person or group to reply to the attack.

Regarding the personal attack rule, the FCC reasoned that the public has a right to hear all sides of a controversial issue but, in the process, individuals and groups should not be maligned without opportunity to reply. Again, no other medium of communication is held to this standard. In 1974, a political candidate took the *Miami Herald* to court arguing that he should have the right to reply to the newspaper's criticism of him. A Florida court upheld the political candidate's right to reply, but the Supreme Court ruled that the decision of the Florida court was unconstitutional and that newspapers were not held to any right of reply.[7] The doctrine of scarcity has always been the rationale for declaring radio and television to be unique media that can be regulated, unlike newspapers.

CANON 35: INSIDE THE COURTROOM

There is another restriction on broadcast programming that affects radio and television, but it is neither a law nor an FCC ruling. The restriction is Canon 35 of the American Bar Association's (ABA) ethical code, which denies to broadcasters the right to take tape recorders or cameras of any kind into the courts. Canon 35 was adopted by the ABA in 1937 when hoards of journalists were flooding the courts and engaging in sensational reporting, causing members of the ABA to fear that individuals were being tried in the press, rather than in the courts.

With Canon 35, two Constitutional guarantees are in direct clash: 1) the First Amendment's guarantee of a free press and 2) the Sixth Amendment's guarantee of a fair trial. Under the Constitution, an individual is guaranteed a

7.   See *The Quill,* July 1974, p. 10.

fair trial, but that guarantee may be nullified by excessive press coverage; on the other hand, in a democracy, court proceedings should be open for examination to insure a fair trial. There is indeed a dilemma involving free press and a fair trial that has not been resolved.

Some states have experimented with allowing some broadcast equipment into the courtroom, but in most U.S. courts, all broadcast equipment is strictly forbidden. Representatives of the electronic news media have vigorously protested this restriction on the flow of information from legal proceedings.

OBSCENITIES, LOTTERIES, LIBEL, AND SLANDER

Other restrictions on programming include laws against programming lotteries, or obscene language, or making libelous or slanderous comments. The law against broadcasting a lottery is written into the Communications Act. Congress defined a lottery as a game involving "prize, chance, and consideration." Concerning the definition of a lottery, one broadcast researcher explained:

> In other words, if you can win something, if you are chosen by luck instead of by skill, and if you had to pay something to get into the game, that's a lottery. If the game involved "the best essay" or "the most beautiful picture," that's something done with skill so it's not a lottery. If you can enter the game completely free, it's not a lottery.[8]

The lottery ruling is very clear and most stations do not have difficulty in interpreting its stipulations. When a station airs or sponsors a contest, it is careful to allow everyone to enter free of charge and, frequently, some skill may be required to win the prize.

Another restriction on programming is the regulation prohibiting obscenities on the air. The statute states that no one shall utter by means of radio communications "obscene, indecent, or profane language." Obscenity is not defined in the statute and this lack has led to problems of interpretation.

Again, other media have greater freedoms than do radio and television, and other media are subject only to checks placed on the media by the consumers. Since the term "obscenity" has not been defined, the concept is, at times, difficult to understand. With changing social values, what may be obscene at one time in history, may not appear obscene at another time. What may be obscene to one group, may not be obscene to another group. What may be obscene at 3:00 PM may not be obscene at 3:00 AM with a different audience. Cable channels, movie channels, pay-TV, and other new forms of home entertainment media have greatly loosened the obscenity rules.

Libel and slander are other areas of concern to broadcasters. Libel is the use of words that "defame, exposing an individual to public hatred, shame, con-

---

8.   Ronald J. Seidle, *Air Time* (Boston: Holbrook Press, 1977), pp. 264–265.

tempt, or ostracism"; libel is the written form while slander is the spoken form. Libelous words usually cause some form of injury to the defamed person— humiliation, loss of job, health problems, and so forth.

Actually, because of the personal attack rule discussed above, libel and slander have not been major problems for broadcasters. However, libel laws are usually written by the individual states, and any radio personality, station owner, or station manager should have at least a cursory understanding of the libel/slander laws of his or her particular state to avoid potential problems.

RULES AND REGULATIONS ON STATION OPERATIONS

Technical Aspects

The vast majority of FCC regulations govern equipment and daily operations. The FCC specifies antenna height, operating power, times of operation, transmitter maintenance, and many other types of very exacting equipment specifications.

To demonstrate the importance of equipment maintenance and station operations, authors Joseph Johnson and Kenneth Jones prepared a list of the FCC rule violations that most often result in some form of penalty from the FCC.[9] Most violations concern engineering practices:

> *Maintenance of equipment log,* including requirement to record quarterly tower-light inspection, failure to enter required weekly antenna base current, etc.
> *Maintenance of transmitter operating log.*
> *Failure to identify the station* by assigned call letters and geographic location at specified intervals.
> *Failure to properly maintain the programming log,* showing all public service announcements, political announcements, commercials, etc.
> *Failure to properly modulate the station's signal.*
> *Failure to maintain proper operating power.*

These types of violations can be decreased by: 1) hiring well-qualified engineers who are familiar with FCC regulations, 2) reading all documents sent to the station from the FCC and filtering these documents to the appropriate station personnel, and 3) developing a daily procedure that includes making all maintenance and operation checks. The engineering staff can protect the station from most FCC violations. But they must perform properly every day, since an FCC representative can make random, surprise visits to any station at any time.

9.   Joseph S. Johnson and Kenneth K. Jones, *Modern Radio Station Practices,* 2d ed. (Belmont, Calif.: Wadsworth, 1978), pp. 225–227.

Licensing

Every broadcast station is licensed by the FCC and each radio station is subject to license renewal every seven years. The seven-year license is a recent development. Prior to 1981, radio and TV stations were licensed for only a three-year period. In the summer of 1981, Congress introduced new legislation that extended the license period for TV stations to five years and for radio stations to seven years.

The license renewal period used to be a very traumatic time for station staff. There were many forms to complete and many facts and figures to send to the FCC. But also in 1981, the FCC adopted a short-form renewal application. This application is barely larger than a postcard. A long renewal form is issued randomly to about 5 percent of the stations. The long form provides the FCC with considerably more information about station operations. Also, the FCC attempts surprise annual inspections of 10 to 15 percent of the radio stations.

When the station license comes up for renewal, the licensee may be challenged by a local group that claims the station does not serve the community. During the 1960s and throughout the 1970s, the number of license challenges increased dramatically due to the demands for media social responsibility by activist and consumer groups.

Most challenges to radio station licenses by activist groups have involved: 1) a concern for employment of women and/or minorities and 2) a concern that the station was not programming to serve the needs of the public. In recent years, the FCC has maintained the belief that the competitive marketplace will force the station to meet programming preferences of the community.

The concerned and careful broadcaster develops a strong public record and will be unafraid of losing a license by the challenge of an activist group. What the broadcaster does fear is the loss of money involved in court and attorneys' fees in fighting a license challenge.

FCC Enforcement

The FCC has ample enforcement power to insure compliance with its rules and regulations. The lowest level of enforcement is the cease-and-desist order that puts the commission on record as having observed a problem and having notified the station to correct the problem immediately.

The second level of enforcement was approved by Congress in 1960— "forfeitures" or fines of up to $1,000 per day of violation, up to a minimum of $10,000 for willful and repeated violations. These first two levels of enforcement—cease-and-desist orders and fines—allow the commission to correct the ways of a station without revoking the station license.

The third level of enforcement is a short-term license renewal. This renewal usually allows the station time to correct its problem before a hearing is scheduled.

The fourth level of enforcement is the FCC's refusal to renew a license, and the fifth level, the ultimate authority of the commission, is to revoke a license before renewal time. Actually, an examination of FCC history reveals that the commission is very reluctant to deny license renewals or to revoke a license. Out of the thousands of broadcast licenses granted and of renewals requested, in the first thirty-six years of the FCC's operation, only 159 licenses have been revoked or denied renewal.[10] The most frequent cause for losing a license is misrepresentations made to the FCC. This usually involves willfully misleading, inaccurate, or false information submitted by the licensee to the commission.

While few broadcasters lose their licenses, most broadcasters find the FCC rules and regulations to be extensive and costly. The costs include:

Expert engineering consultants to insure station compliance with all of the technical standards

Personnel to monitor community needs and concerns and file FCC reports

An attorney to assist in interpreting FCC rules and regulations

Personnel to type, file, and maintain records

Paper, postage, telephones, and other supplies and equipment needed to reply to FCC requests.

Because the industry has been concerned about overregulation from the federal government, broadcasters have tried to regulate themselves to prevent further action by government. This is the next part of the regulatory process to be examined.

INDUSTRY REGULATIONS

The National Association of Broadcasters (NAB) was formed in 1923 as a voluntary effort on the part of the industry to regulate itself. In an effort to prevent further governmental regulations, the NAB adopted a Code of Standards for radio in 1929. The Radio Code has been revised over the years.

Until 1976 stations could join the NAB without subscribing to the code. However, since the code specified time standards for radio advertising, rules governing the acceptability of particular types of ads, as well as appropriate advertising practices, many radio stations did not subscribe to the code. If a station subscribing to the code was found in violation, the NAB could withdraw permission for the station to display the NAB Code of Good Practices.

In addition to its regulating broadcasting, the NAB is the primary lobbying agency for the interests of commercial broadcasters. The association regularly informs broadcasters of bills before Congress, solicits opinions from the broadcasters, and launches campaigns to insure that the broadcasters' voices are

10. John A. Abel, Charles Clift III, and Fredric A. Weiss, "Station License Revocation and Denials of Renewal, 1934–1969," *Journal of Broadcasting* 14 (Fall 1970): 411–421.

heard. In recent years, the NAB has expressed the industry's opinions on: the Communications Act rewrite, new technological standards for radio, license renewal practices, the fairness doctrine, limitations on advertising, and many other issues.

Over the years the NAB has grown stronger through increased membership and increased revenues. It represents well the interests of commercial broadcasters, but there are several problems with its claim as a regulatory agency. There may be a conflict of interest with an industry-lobbying association regulating itself in the best interests of the general public. Also, not all stations subscribe to the NAB and its regulatory code. If a broadcast station is in violation of NAB standards, the NAB enforcement power is limited and almost non-threatening. The NAB is the major industry association, however, and probably will continue to increase its strength in the industry.

Other industry organizations include the Radio, Television News Directors Association (RTNDA), and the National Radio Broadcasters Association (NRBA). These organizations function on behalf of the industry and also challenge the industry to high standards.

The NAEB is the organization that represents the noncommercial public broadcasters. The NAEB performs similar functions for its members as does the NAB for its members. The NAEB is a much smaller organization and functions on much smaller financial resources than does the NAB.

The NRBA is a relatively new organization and is at times in competition with the NAB for membership. The NRBA represents the interests of radio station owners and operators, but it is a much smaller organization than is the NAB and it too, functions on much smaller financial resources.

In addition to the various organizations and associations, the individual networks and stations have codes that they have developed for their performance. The networks screen advertisements and have program standards departments to regulate broadcast materials. Most stations have a defined station policy governing acceptable advertising and programming performance. Even with these efforts, the primary source of regulations is still the various governmental agencies that are specially charged with particular regulatory tasks.

THE PUBLIC

The general public also has both a direct and an indirect impact on regulating the mass media. One of the most direct roles of the public is that of consumer. If the public does not like a radio format, it will not listen to the station, so the format is destined to change. This is a free-market form of regulating the industry by consumer practices.

Another role that individuals serve is in letting the station know what they want to hear. One letter to a station makes a big impression on the station personnel. At many stations, letters from individuals in the listening audience are posted on the bulletin board for all staff members to see. The management takes action when many calls or letters suggest a change in programming. After all, the station operates to gain a large audience, so of course the management wants good public relations to insure large audiences and high rates for advertising time.

If individuals are not satisfied with the response they receive from a radio station or a radio network, they can correspond with the FCC in an effort to bring about a desired change. In recent years, individuals with similar goals have united into citizen groups that organize campaigns to change media performance. The campaigns usually include visits to the station to talk with the management and suggest change. If this avenue of action is not successful, the group may file a petition with the FCC requesting that the FCC intervene.

REGULATORY FORCES IN THE UNITED STATES

Broadcasting is, indeed, the most regulated of our mass media. Radio is regulated directly and indirectly by the FCC, Congress, the courts, other governmental agencies, the industry itself, and the general public. Exhibit 9.1 shows the regulatory forces on the broadcaster.

In this chapter, we have already discussed most of the regulatory forces, but some of the regulatory agencies may need clarification. The White House, for example, regulates broadcasting by appointing FCC commissioners and by making its position known on issues facing the broadcasting industry. The Congress has its impact on broadcasting, since the Senate must approve FCC commissioner appointments and numerous congressional committees are involved in communication issues. Just two of the important congressional committees are the Senate Communications Subcommittee and the House Communications Subcommittee. Also, any changes in the Communications Act of 1934 must be approved by Congress.

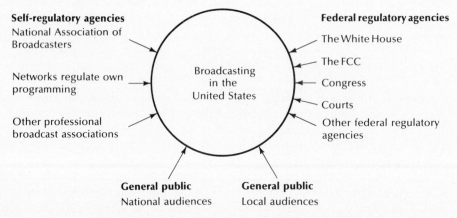

Exhibit 9.1.   The Regulatory Forces on Broadcasting

The courts regulate broadcasting by reviewing FCC decisions that are legally challenged. The FCC is the primary broadcasting regulator, but numerous other federal agencies have direct and indirect influence. These agencies include: Federal Trade Commission, Internal Revenue Service, Department of Justice, Small Business Administration, and others.

There is always an adversary relationship between the broadcaster and regulatory forces. An adversary relationship is healthy in a democracy and insures that the industry will not be underregulated or overregulated. Broadcast regulations and regulatory forces change from time to time. New issues are constantly raised and resolved. The broadcasting student should carefully monitor these issues and follow the various arguments. The broadcasting trade publications are the information sources monitoring the regulatory forces.

BIBLIOGRAPHY

Broadcasting magazine, 1983.

Broadcasting Yearbook, 1983.

Cole, Barry, and Oettinger, Mal. *Reluctant Regulators: The FCC and the Broadcast Audience.* Reading, Mass.: Addison-Wesley, 1978.

Dennis, Everette E., and Merrill, John C. *Basic Issues in Mass Communication.* New York: Macmillan, 1984. See Chapter 1, "Freedom of Press" and Chapter 2, "Media–Government Relationship."

Kahn, Frank J., ed. *Documents of American Broadcasting.* New York: Appleton-Century-Crofts, 1968.

Kittross, John M., and Harwood, Kenneth, eds. *Free and Fair: Courtroom Access and the Fairness Doctrine.* Temple University: Association for Professional Broadcasting Education, 1970.

Rubin, Bernard, *Media, Politics, and Democracy.* New York: Oxford University Press, 1977.

QUESTIONS FOR REVIEW AND DISCUSSION

1. Why is broadcasting more heavily regulated than any other communication medium in this country? Do you believe the regulation of broadcasting is justifiable?

2. How is the FCC appointed and what are its specific duties as they relate to radio broadcasting?

3. Why is the Communications Act of 1934 of such significance to broadcasters? Do you believe that the act is outdated? If so, why?

4. What is Section 315 of the Communications Act of 1934? Is Section 315 a good ruling?

5. What is the fairness doctrine? Provide some specific examples of how the fairness doctrine functions?

6. What is the personal attack rule? Why must broadcasters be especially aware of this regulation?

7. For what purpose was Canon 35 instituted? Do you believe that Canon 35 serves its original purpose? What two constitutional guarantees are in conflict in the ruling?

8. For broadcast purposes, what is the definition of a lottery?

9. What are the FCC regulations on obscenities? Why are obscenities difficult to define?

10. What is the difference between libel and slander?

11. Does the FCC frequently revoke station licenses? On what grounds is the FCC most likely to revoke a station license?

12. Why are the broadcast media subject to complaints from the listening or viewing public? Do you believe that station license challenges improve the quality of broadcast programming?

13. What specific powers does the FCC have to regulate broadcasting? List these from the strongest enforcement penalty to the weakest.

14. How does the general public indirectly regulate broadcasting?

15. List the various agencies and institutions that directly and indirectly regulate broadcasting.

16. If you were writing a new communications act, what new types of legislation would you include in the act? What would you leave out of the old act?

———————————————————————

——————————————— *Chapter Ten* ———————————

———————————————————————

RADIO: THE FUTURE

*The broadcast station of the future will be alive with activity.
The activity will be generated by satellites, digital audio systems, and computers, and by a few operators who know how
to interact with the new technology.*

The radio industry has experienced tremendous growth in the last twenty years.
Fifteen years ago 100 schools awarded 1,500 undergraduate degrees in broadcasting; now, over 8,600 students graduate yearly with undergraduate degrees
from 205 colleges and universities. The National Association of Broadcasters
reports that radio provides most of the entry-level professional positions in
broadcasting and accounts for more than 116,000 full-time jobs that are now
available in the industry.[1] What does the future hold for the broadcast graduate
seeking a position in the radio industry? This chapter explores the future of
radio as it relates to the career goals of today's college graduates.

INDUSTRY FACTORS

The Competition

There are nearly 9,000 radio stations operating in the United States. Of these
4,600 are commercial AMs, over 3,000 are commercial FMs, and over 1,000 are
noncommercial FMs. There are an estimated 425 million radio sets in the
United States with an estimated 310 million sets in the home and 115 million
sets outside the home.[2]

1. *Radio in 1985,* The National Association of Broadcasters, 1977, p. 8.
2. *Broadcasting Yearbook,* 1983.

All indications are that radio is strong, especially at the local level. Radio serves local interests by providing the community with information of interest to local people. Local advertisers who could not afford to advertise on television can afford local radio. But all signs are not rosy. There are some good indicators for radio and there are some danger signs, and some of the good indicators are, at the same time, caution indicators.

One good indicator for radio has been the growth of radio stations. In addition to the many stations currently on the air, the FCC recently authorized FM space that could add an additional 500 to 1,000 more stations.[3] While the growth of stations appears to be a strong indicator for radio, it has not necessarily brought strength to the industry. What has happened as a result of the increased number of stations is that competition has become extremely keen.

The radio broadcaster is competing for audience time with not just every other radio station but with every other mass medium as well. In many communities an individual may have eight or nine radio stations, ten or more television stations (including cable), two local newspapers, as well as movies and a vast assortment of magazines. What this means is that the radio audience has been segmented. This situation has forced many stations to rely on the least expensive operation possible. Many stations have become automated. A computer can be programmed to play recorded music at less cost than highly trained radio operators can broadcast live programming. The result is that many radio jobs will no longer hold the interest of a college graduate.

The Audience

The good projections for radio are definitely for radio's audience. A National Association of Broadcasters' publication makes the following predictions about radio's audience.[4]

The number of households will increase at a faster rate than the total population. This means more households, more products purchased. NAB writes: "It can be plausibly argued that six people living within the same household will purchase different (and probably fewer) items than the same six people living apart. The more households, the more potential buyers and the greater the need for advertisers' products and services." More advertising means that every radio station has a chance to increase its advertising revenues.

The proportion of blacks in the total population will increase. In 1975, blacks comprised 11.5 percent of the population; by 1985, they will comprise 12.1 percent of the U.S. population. "Radio stations serving the minority market will be doing proportionately better economically...."

Women will continue to outnumber men. "As is currently the case females will make up a majority of the total population in 1985. A good deal of programming and advertising on radio is currently directed explicitly toward

3. *Broadcasting,* May 30, 1983, p. 31.
4. *Radio in 1985,* The National Association of Broadcasters.

female listeners. There is every indication that such programming and advertising slants will continue."

The largest portion of the population will be at the peak spending age. "Forty percent of the total U.S. population in 1985 will be between twenty-five and fifty-four years old. People in the twenty-five to fifty-four age group are the people buying homes, purchasing cars, acquiring new furniture, spending on children, making luxury item purchases." With 40 percent of the total U.S. population in the peak spending age, there should be a greater need to advertise goods and services for these buyers. This should lead to more advertising dollars for radio.

Americans will be better educated than ever before. In 1985 Americans twenty-five and over will be achieving the highest educational levels in history: 30 percent of this group will have graduated from or attended college. (This figure was only 17 percent in 1960.) Among the college trained, 50 percent are in upper income earning brackets. Better educated listeners mean better paid listeners; better paid listeners have more money to spend on discretionary items. Again, this situation could lead to more advertising for radio.

Average family income will increase dramatically. "The number of families with incomes of $24,000 or more (in constant 1975 dollars) will more than double in 1985. All of this means more discretionary spending. . . . Newly developed tastes could produce new sources of radio advertising."

Women will make up 40 percent of the total U.S. labor force. "This trend has produced, and will continue to produce, more diversified family and individual spending due to increased discretionary family income. Advertising geared to women need not be solely related to 'household' goods or 'homemakers' items." The working woman is producing new markets for a wide variety of goods and services—everything from fast foods to microwave ovens to expensive clothing.

The population will continue to be extremely mobile. "In ten years' time audiences should be bigger than ever before, since many suburban dwellers will continue to work in the city. The movement of people out of state or away from the community means radio stations can no longer rely on 'old friends' and 'loyal listeners' to keep things going over the long stretch. The effort to woo new listeners must be a continuous one." The need to woo new audiences and maintain old audiences will mean that the successful station must remain alert to the needs and interests of the changing audience. Listeners will go to the stations that provide programming attuned to satisfying their unique interests.

What do these projections mean for the broadcast graduate? They mean that radio has a viable future. They mean that the audience is potentially strong for radio, but the audience will use the medium only if radio provides a unique service. Throughout the 1980s and 1990s, radio could be a creative medium with each station having unique characteristics, if creative management leads the way. The potential for a strong radio industry is obvious, but radio will not realize its potential without the input from today's graduates who have a vision of what the industry could be.

TECHNOLOGY

Satellites

Satellites are the newest and the biggest story in radio technology. Satellite communication has raised the engineering standards in the radio industry. Due to the increased bandwidth of the carrier wave, much higher music quality signals can be modulated on the carrier wave and modulation techniques via satellite are less prone to noise and distortion than are conventional broadcast techniques.

Not only are satellites introducing higher technical standards in the radio industry, but satellites are also responsible for the introduction of numerous new radio networks. The satellite technology allows for many more channels to be broadcast simultaneously. This aspect of satellite broadcasting is responsible for the development of many new programming options in the radio industry.

The RKO radio network has changed totally to a satellite distribution system. Mutual Broadcasting is converting its transmission system to a satellite system. ABC, CBS, and NBC radio networks are distributing some programming via satellite. Numerous new radio networks are being formed that will use all satellite distribution systems. The networks are providing the stations with earth reception systems or, in some cases, are leasing the receiving systems to the stations.

The new satellite networks are offering many stations a higher quality of programming than the local station would be able to provide its listeners—especially in the late night and early morning hours. Most of the satellite networks are providing twenty-four-hour programming, which will allow stations to operate for more hours at lower costs. The satellite networks will eliminate the necessity of having on-air personalities at the local station during weekends, late nights, and early mornings. The new networks will offer a wide variety of programming options to the local audience.

This new technology promises improvements in programming and in engineering but again, the picture is not altogether rosy. The small market station may not be able to afford this new technology. The costs of keeping up with the technological changes in radio may ultimately put the small broadcaster out of the radio industry, while failure to keep up with the new technology may cause the small market broadcaster to lose the local audience to the larger stations in the nearby cities. New technology may introduce great profits and a bright future for some stations and may be the death knell for others.

Microprocessors in Radio

Computers have entered the radio industry and have introduced many changes in station operations. The computer is responsible for the automated station operations that were discussed in previous chapters. But the computer will continue to offer the station operator more automated programming systems. New audio consoles are now available that can be programmed to open particular channels at precise times. The new microprocessor-controlled boards allow a signal to be mixed in a channel and the signal level to be stored in memory for future use. The audio consoles of the future may be so automated as to require only a minimum of manual maneuver.

After a recent broadcast equipment exposition, the broadcasters were asked what they would like to see in future broadcast technology. One broadcaster described the system that he would ultimately like to see at his station. He said that he wanted an automated system that would allow him to:

> take the entire library . . . typically 4,000 cartridges of music selections, put them into a computer, enter this whole library in the computer in digital form. Your console can be operated by a microprocessor, your transmitter operated by microprocessor, and you can have the person sit there, dial up your music, program the entire station, look at your technical facility and have that all computerized.[5]

The broadcast station of the future may be alive with activity, but with few humans producing the activity. The activity will be generated by the computer and by a few operators who know how to interact with the new technology.

What does this new technology mean to the broadcast student? It introduces many job options in the technical fields, but it also means that today's broadcasting students must know something about the computer. The student must know how the computer functions and must know how to interact with the computer. There is absolutely no question that computers will play a bigger and bigger part in the radio industry of tomorrow.

FUTURE RADIO TECHNOLOGY

Digital Recording Systems

Traditional recording methods require a sound signal to be converted into an electrical signal that is stored on magnetic audio tape. The electrical signal pattern stored on the tape is then converted back into an audio signal for playback. Digital recording is a method of converting a continuous sound signal into a sequence of numerical values that are stored on magnetic tape discs. The numerical values are the various amplitudes of the audio signal being recorded.

In digital recording, the tape recorder samples the amplitude of the audio signal 40,000 times a second at intervals set by the sampling frequency. Exhibit 10.1 contrasts the recording method in traditional recording and the recording method used in digital audio.

Digital audio provides some distinct advantages over traditional recording: 1) digital audio will provide better dynamic range in a recorded signal than will traditional recording methods, 2) digital recording, unlike traditional recording,

---

5. *Broadcasting,* April 27, 1981, p. 61.

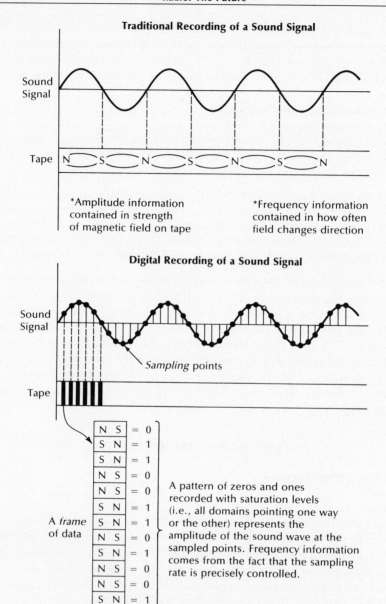

Traditional Recording of a Sound Signal

Sound Signal

Tape

*Amplitude information contained in strength of magnetic field on tape

*Frequency information contained in how often field changes direction

Digital Recording of a Sound Signal

Sound Signal

Sampling points

Tape

N	S	= 0
S	N	= 1
S	N	= 1
N	S	= 0
N	S	= 0
S	N	= 1
S	N	= 1
N	S	= 0
S	N	= 1
N	S	= 0
N	S	= 0
S	N	= 1
S	N	= 1
N	S	= 0

A *frame* of data

A pattern of zeros and ones recorded with saturation levels (i.e., all domains pointing one way or the other) represents the amplitude of the sound wave at the sampled points. Frequency information comes from the fact that the sampling rate is precisely controlled.

Exhibit 10.1.   Traditional vs. Digital Recording Methods

virtually eliminates distortion, 3) with digital audio there is no need for a noise reduction system since the digital recording has no hiss or white noise, 4) using the digital recording system, any number of copies of the original signal can be made with no loss of quality.

While digital audio recorders are currently being manufactured by a number of companies including Sony, 3M, Mitsubishi, and Ampex, the use of digital audio in the radio industry is some years into the future. The reason for this is primarily that standards have not been set within the audio industry for the digital systems. Thus, equipment manufactured by one company is not compatible with equipment manufactured by another.

Complete acceptance of digital audio in the radio industry depends on four major factors:

1. Digital systems must be standardized.
2. The costs of the digital systems must be reduced so that broadcasters can afford to convert from the traditional systems to the digital systems.
3. The Federal Communications Commission must approve a standardized digital system for broadcast use.
4. The demand must be present from the listening public for the higher quality signal that digital audio can provide.

The four factors listed above mean that we will not see a rush to digital audio in the radio industry. The conversion from traditional audio to digital audio will take place over the next five to fifteen years. Emil Torick, director of audio development for the CBS Technology Center, has described what he envisions as a complete digital studio:

> Whether in a broadcast station or a recording plant, there will be huge gains in fidelity, precision, and accuracy of control when the signal stays in digital form throughout its manipulation within the plant. Everything done there—amplitude control, modulation control, filtering, equalization, processing, etc.—is far better done with a digital signal. There should be no necessity to convert back into analog form until the signal leaves the plant. And even that conversion will be eliminated when the signal on the air is digital, as it may be one day.[6]

Digital Audio Discs (DADs)

In addition to digital tape recording, digital techniques have also invaded the disc recording industry. The traditional long-playing record or disc measures approximately twelve inches in diameter, is plastic, and contains a series of grooves that represent the electrical signal that is converted into an audio signal on the record player. The new digital discs are composed of a layer of

6. "Digital Audio: A View of Today and Tomorrow," *Broadcast Management/Engineering* (BME) 18:2 (February 1982), p. 46.

metal on which is encoded a laser signal. The disc is then encased in a layer of plastic. The digital audio disc measures only 4.5 inches in diameter.

Like other digital audio systems, the digital discs have advantages over traditional audio discs:

1. The reproduced sound of the digital disc is far superior in fidelity to the sound quality of traditional discs.
2. The digital discs are much smaller than traditional discs and are easier to store.
3. The digital discs are protected in a thin layer of plastic so that irregularities and scratches do not influence the playback signal.
4. The digital discs are not affected by dust or by finger grease, making them more durable and easier to handle.

The digital discs must be played on special playback systems that use a laser beam to decode the signal. These new digital audio players are being manufactured by a variety of companies including Sony, Philips, Studer Revox, Pioneer, Matsushita, Crown, Dual, and Nakamichi. The digital disc players are currently expensive, but as the technology becomes more standard, as demand increases, and competition increases within the industry, the cost of the disc players will decrease.

The digital discs may ultimately have a major impact on the radio industry. One industry publication has predicted:

> The DAD (digital audio disc) will have an enormous impact on the consumer high fidelity market, elevating listener expectations as the machines get into the hands of a sizeable number of users. On that score alone the coming of the DAD will be a crucial event for radio broadcasters. As the DAD reaches a widening public, the radio broadcaster more and more will have to look to the technical quality of the signal on the air.[7]

Indeed digital audio systems may make a major impact on the radio industry. The systems represent new technology that must be evaluated by professionals in the radio industry.

Technology and the Broadcast Student

With satellites, computers, and digital audio, the radio industry is undergoing a technological revolution. The serious broadcasting student must follow the impact of the new technology on radio broadcasting and must be prepared to cope with it on the job. This means that the student must read the professional

7. "Digital Audio Discs Glimmer on the Horizon," *Broadcast Management/Engineering* (BME) 18:2 (February 1982), p. 49.

publications in the broadcasting industry and must take a curriculum that will prepare him or her to use the new equipment that is likely to be found on the job.

CAREER OPPORTUNITIES IN BROADCASTING

Not only has a revolution taken place in broadcast technology, but also a revolution of sorts has taken place in the broadcast labor market. There used to be just a few job titles—station manager, program director, news director, DJ, and so on. Now there are as many job possibilities as there are creative persons: system and facility designers, broadcast insurance agents, news specialists, media time buyers, and many others that were unheard of just a few years ago.

Today's broadcast graduates should prepare themselves not only for the jobs in radio production, sales, promotion, and management but also for the wide variety of jobs that has grown in and around the broadcasting industry. A survey of *Broadcasting Yearbook* lists the following kinds of job roles in broadcasting:[8]

Advertising agents	Broadcast insurance agents
Brokers	Public relations and promotion specialists
Program syndicators	Broadcast attorneys
Station representatives	Talent agents
Management consultants	Broadcast employment agents
News consultants	Research specialists
Market consultants	News specialists
Engineering consultants	Programming distributors
Programming consultants	Media time buyers
Market and audience researchers	Media coordinators
System and facility designers	

The above list isn't complete, but it does show the wide variety of opportunities that are available in the radio industry. Too frequently students think of only the obvious jobs, but many opportunities are available for the creative thinker. *Broadcasting Yearbook* lists 128 consulting firms that are prepared to advise stations on everything from ascertainment, to equal employment practices, to building an audience, to programming, to station management practices. The same yearbook lists seventy-two research firms that offer services from market research, to station image studies, to rate card and format studies. There are also a wide variety of radio programming services that offer everything from a complete format, to a music library, to station IDs.

Even though there are many services available, there is always room for the innovator. These services did not just happen, they developed as the result of a

8. *Broadcasting Yearbook,* 1983.

need in the industry that someone saw and had the expertise to fill. Today's broadcasting graduate should spend some time in the library examining such publications as *Broadcasting Yearbook, Broadcasting* magazine, *Advertising Age* and many other publications that indicate the present opportunities in the communications industries and the trends in the mass media. The future belongs to the individual who knows where the industry is headed and is prepared to meet the needs of the industry.

PREPARATION FOR JOB HUNTING: THE RÉSUMÉ

The first item a student will need for job hunting is a résumé. The résumé provides the employer with essential information about the potential employee—age, educational background, special skills, previous jobs, and specialized job experience. The résumé should "put the best foot forward"; it should be well organized, neat, and attractive. The student may want to utilize the services of a printing company that offers quality printing and a variety of paper stocks.

In preparation for writing the résumé, the student should spend part of his or her college career training for the future—preparing broadcast materials (television and radio programs, films, instructional materials), serving internships, or working part-time at a local television station, radio station, advertising agency, or other broadcast-related business. On the résumé, the student should list the relevant broadcast courses she or he has taken, the various media materials the student has produced, and all relevant job experience. The résumé should be prepared in consultation with the college placement office or with a faculty adviser.

The key to landing a good job upon graduation is preparing for that job while in college. The careful student doesn't wait until the senior year to begin preparation. Such a student creates opportunities and builds a strong résumé through selection of a meaningful curriculum, preparation of media materials, and employment in the communications industries during each year of college. The student who keeps career aspirations in mind will be easily prepared for the task of job hunting in the senior year.

THE FUTURE OF THE RADIO INDUSTRY

Projections indicate that the total listening audience for radio will continue to grow. Just since 1969, the audience has grown 25 percent. Pundits make the following predictions:

> The average quarter-hour American radio listening audience (all persons twelve plus years of age) will increase from 29 million in 1976 to 34 million by 1985. These men and women and teens will be spending three hours fifteen minutes a day with radio.

The operating profit margin of radio stations is expected to jump from 13 percent to 16.2 percent. Every analysis shows not only continued good health, but improved health within the industry. These projections for radio apply across the board in every section of the country in every size market.[9]

The future of radio looks bright, but the student should remember that the competition will remain keen both within the industry itself and among broadcasting graduates competing for the best jobs. The number of broadcasting graduates will continue to grow in the near future. This means that college graduates must be well prepared. If radio is to develop to its full potential, the industry must have well-trained employees who thoroughly understand the industry and are prepared to undertake creative experiments in its further development.

BIBLIOGRAPHY

Advertising Age. A weekly publication.

Bittner, John R. *Professional Broadcasting.* Englewood Cliffs, N.J.: Prentice-Hall, 1981.

Bittner, John R. *Professional Broadcasting: A Brief Introduction.* Englewood Cliffs, N.J.: Prentice-Hall, 1981.

Broadcasting magazine.

Broadcast Management/Engineering (BME).

Broadcasting Yearbook, 1983.

Ellis, Elmo I. *Opportunities in Broadcasting.* Skokie, Ill.: National Textbook, 1977.

Gross, Lynn Schafer. *The Internship Experience.* Belmont, Calif.: Wadsworth, 1981.

Radio Active. A magazine published by the National Association of Broadcasters (Broadcasting Department), 1171 N Street, N.W., Washington, D.C. 20036. For members of the NAB.

QUESTIONS FOR REVIEW AND DISCUSSION

1. What are the good indicators for the radio industry and what are the danger signs?
2. What are some of the projections for the radio audience?
3. What changes have satellites introduced in the radio industry?
4. How have computers changed radio station operations? How will they continue to change future operations?
5. What are the advantages of digital recording systems over traditional recording systems?
6. What factors will determine the acceptance of digital audio systems in the radio industry?
7. What are the advantages of digital discs over traditional discs?

9. *Radio in 1985,* The National Association of Broadcasters.

8. What can the broadcasting student do to prepare for the challenges introduced by the new technology?

9. What opportunities are available to the radio broadcaster outside the commercial radio station?

10. What type of information should be included in a résumé?

11. What are the predictions for radio's future?

THE METRIC SYSTEM

Several years ago, Congress instituted the metrification process whereby the U.S. will gradually transform all measurements to the metric system. This is the system used by most of the rest of the world now and most imported products already use it fully. The U.S. auto companies have begun to use metric units, and many other domestic manufacturers are also switching. You now routinely see road markers with both mile and kilometer units, and frequently temperatures are displayed in both Fahrenheit and Celsius (centigrade) units. The actual conversions for some of the more useful quantities are:

Quantity	Customary Unit	Metric Equivalent
Length	1 foot	0.305 meters
Length	1 yard	0.915 meters
Length	1 mile	1.61 kilometers
Mass (weight)	1 pound	0.45 kilograms
Mass (weight)	1 ounce	28 grams
Temperature	1 degree Fahrenheit	0.55 degree Celsius
Temperature	Freezing point of water (32°F)	0°C
Temperature	Boiling point of water (212°F)	100°C
Liquid Volume	1 quart	0.95 liter
Liquid Volume	1 ounce	30 milliliters

These are most of the "everyday" units; there are many others in the metric system but most of them involve multiples of or fractions of the basic units. One nice aspect of the metric system is the ability to make larger or smaller units by use of prefixes based on multiples of ten. For example, instead of dealing with miles, which are 5,280 feet, we use kilometers, which are 1,000 meters; and instead of using inches (one-twelfth of a foot), and sixteenths of an inch, we use centimeters (one-hundredths of a meter) and millimeters (one-thousandths of a meter). Thus all the units of length, for example, are related by multiplying or dividing by multiples of 10 (i.e., 10, 100, 1,000, etc.). The possible prefixes and their meanings are:

Prefix	Meaning
micro-	one-millionth
milli-	one-thousandth
centi-	one-hundredth
deci-	one-tenth
deca-	ten
hecto-	hundred
kilo-	thousand
mega-	million

Examples of use of the prefixes include the *centi*meter, which is one-hundredth of a meter and is roughly comparable to the inch measure (2.54 centimeters = 1 inch), and the *kilo*gram, which is one thousand grams (equivalent to 2.2 pounds).

Conversion of the commonly used units from customary to metric and vice versa is shown in the accompanying table.

Approximate conversions from customary to metric and vice versa

	When you know:	*You can find:*	*If you multiply by:*
LENGTH	inches	millimeters	25
	feet	centimeters	30
	yards	meters	0.9
	miles	kilometers	1.6
	millimeters	inches	0.04
	centimeters	inches	0.4
	meters	yards	1.1
	kilometers	miles	0.6
AREA	square inches	square centimeters	6.5
	square feet	square meters	0.09
	square yards	square meters	0.8
	square miles	square kilometers	2.6
	acres	square hectometers (hectares)	0.4
	square centimeters	square inches	0.16
	square meters	square yards	1.2
	square kilometers	square miles	0.4
	square hectometers (hectares)	acres	2.5
MASS	ounces	grams	28
	pounds	kilograms	0.45
	short tons	megagrams (metric tons)	0.9
	grams	ounces	0.035
	kilograms	pounds	2.2
	megagrams (metric tons)	short tons	1.1
LIQUID VOLUME	ounces	milliliters	30
	pints	liters	0.47
	quarts	liters	0.95
	gallons	liters	3.8
	milliliters	ounces	0.034
	liters	pints	2.1
	liters	quarts	1.06
	liters	gallons	0.26
TEMPERATURE	degrees Fahrenheit	degrees centigrade	5/9 (after subtracting 32)
	degrees centigrade	degrees Fahrenheit	9/5 (then add 32)

ALL ABOUT DECIBELS

The decibel unit, abbreviated dB, allows us to express very large differences in amplitude or power levels with relatively small, more manageable numbers. The actual definitions are as follows:

(1) *Amplitude equation:*
Difference between two amplitude levels in dB
$$= 20 \log(A_2/A_1),$$
where *log* is the logarithm to the base 10, the so-called common logarithm. The logarithm is the power or exponent of 10 which gives the number desired. For example,
$$\log(10) = 1 \text{ since } 10^1 = 10,$$
$$\log(100) = 2 \text{ since } 10^2 = 100,$$
$$\text{and } \log(1000) = 3 \text{ since } 10^3 = 1000, \text{ etc.}$$
(2) *Power or Intensity equation:*
Difference between two power or intensity levels in dB
$$= 10 \log(P_2/P_1).$$

A few examples will serve to clarify these definitions. The amplitude of an audio signal is given by its voltage. If an amplifier produces an output voltage of 100 volts for an input of only 1 volt, then the voltage gain can be expressed in decibels as

$$\text{Gain in db} = 20 \log (V_{out}/V_{in}) = 20 \log(100/1)$$
$$= 20 \log(10^2) = 20 \times 2 = 40 \, dB.$$

Suppose an amplifier has a maximum power output of 15 watts and another has a maximum power output of 150 watts. The difference between these two amplifiers' output levels in decibels is

$$\text{Difference in Power in dB} = 10 \ \log(P_2/P_1) = 10 \ \log\left(\frac{150 \ \text{watts}}{15 \ \text{watts}}\right)$$
$$= 10 \ \log(10) = 10 \times 1 = 10 \ dB.$$

There is one more complication to all of this. Sometimes we ask the question in such a way that the difference in decibels comes out negative. For example, if a microphone has an output voltage of a thousandth of a volt (a *milli*volt), the *sensitivity* of the microphone is obtained by referring this output voltage to a *standard* or *reference* voltage of 1 volt. This can be expressed as

$$\text{Sensitivity of microphone in dB} = 20 \ \log\left(\frac{.001 \ \text{volt}}{1 \ \text{volt}}\right)$$
$$= 20 \ \log(10^{-3}/1) = 20 \ \log(10^{-3})$$
$$= 20 \times (-3) = -60 \ dB.$$

Note that this result depends on the fact that:

$$\text{one-thousandth} = 1/1000 = .001 = 1/10^3 = 10^{-3},$$
and that: $\log(10^{-3}) = -3$ since the exponent is -3 in this case.

One rarely needs to do calculations like those above in the actual use of the radio studio and its associated equipment. The purpose of this appendix is to show how it is done and to indicate that quantities expressed in decibel units always involve a difference between two levels; a signal cannot just be this or that many decibels, it must be this or that many decibels with respect to another signal.

Several more points are worth mentioning about decibels. It is interesting to relate the defining equations for power levels and amplitude levels. In general, the power content of a sound wave or its electrical equivalent, the audio signal, depends on the square of the amplitude. That is, if A is the amplitude, then the power depends on A^2. If we have two signals with amplitudes $A_1 = 10$ volts and $A_2 = 100$ volts, for example, then the *difference in dB* $= 20 \log(A_2/A_1) = 20 \log(100/10) = 20\log(10) = 20 \ dB$. For the same voltages, the powers would be (since power depends on A^2) $P_1 = A_1{}^2 = 10^2 = 100$ and $P_2 = A_2{}^2 = 100^2 =$

10000. Then, in terms of power levels, *the difference in dB* = $10 \log(P_2/P_1)$ = 10 log (10000/100) = 10log(100) = *20 dB*. This calculation shows that whether one knows the relative power levels *or* the relative amplitude levels, the difference in decibels is the same.

In working with VU meters, which read out directly in decibels, it is useful to remember that an increase of 10 dB is equivalent to a doubling of the sound "loudness." That is, we hear a 10 dB increase as about twice as loud. This means, for example, that an amplifier producing 150 watts of output power will sound only about twice as loud as one producing 15 watts.

INDEX